Visible and Ultraviolet Spectroscopy

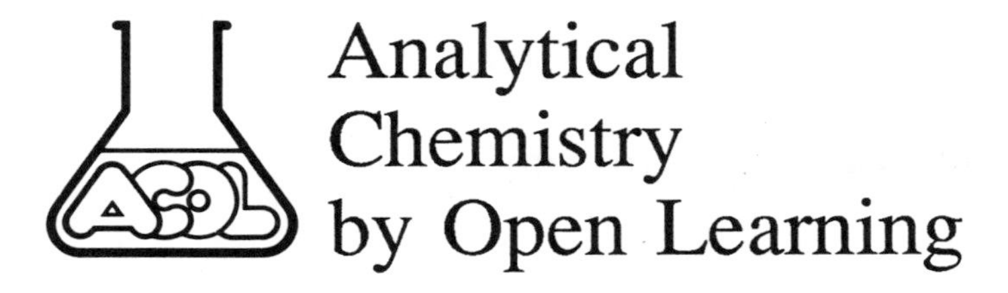

Project Director
BRIAN R CURRELL
Thames Polytechnic

Project Manager
JOHN W JAMES
Consultant

Project Advisors
ANTHONY D ASHMORE
Royal Society of Chemistry

DAVE W PARK
Consultant

Administrative Editor
NORMA CHADWICK
Thames Polytechnic

Titles in Series:

Samples and Standards
Sample Pretreatment and Separation
Classical Methods
Measurement, Statistics and Computation
Using Literature
Instrumentation
Chromatographic Separations
Gas Chromatography
High Performance Liquid Chromatography
Electrophoresis
Thin Layer Chromatography
Visible and Ultraviolet Spectroscopy
Fluorescence and Phosphorescence Spectroscopy
Infrared Spectroscopy
Atomic Absorption and Emission Spectroscopy
Nuclear Magnetic Resonance Spectroscopy
X-ray Methods
Mass Spectrometry
Scanning Electron Microscopy and X-ray Microanalysis
Principles of Electroanalytical Methods
Potentiometry and Ion Selective Electrodes
Polarography and other Voltammetric Methods
Radiochemical Methods
Clinical Specimens
Diagnostic Enzymology
Quantitative Bioassay
Assessment and Control of Biochemical Methods
Thermal Methods
Microprocessor Applications

Visible and Ultraviolet Spectroscopy

Analytical Chemistry by Open Learning

Authors:
RONALD C DENNEY
Thames Polytechnic

ROY SINCLAIR
Paisley College of Technology

Editor:
DAVID MOWTHORPE

on behalf of ACOL

Published on behalf of ACOL, Thames Polytechnic, London
by
JOHN WILEY & SONS
Chichester · New York · Brisbane · Toronto · Singapore

Published by permission of the Controller of
Her Majesty's Stationery Office

Library of Congress Cataloging in Publication Data:

Denney, Ronald C.
Visible and ultraviolet spectroscopy.

(Analytical Chemistry by Open Learning)
Bibliography: p.
1. Absorption spectra—Programmed instruction.
2. Ultraviolet spectroscopy—Programmed instruction.
3. Chemistry, Analytic—Programmed instruction.
I. Sinclair, Roy. II. Mowthorpe, David.
III. ACOL (Project) IV. Title. V. Series: Analytical Chemistry by Open Learning (Series)
QD96.A2S56 1987 543'.0858 87-25243
ISBN 0 471 91378 2
ISBN 0 471 91379 0 (pbk.)

British Library Cataloguing in Publication Data:

Denney, Ronald
Visible and ultraviolet spectroscopy.—
(Analytical chemistry).
1. Visible spectroscopy 2. Ultraviolet spectroscopy
I. Title II. Sinclair, Roy
III. Mowthorpe, David IV. ACOL V. Series
535.8'43 QC454.V/

ISBN 0 471 91378 2
ISBN 0 471 91379 0 Pbk

Printed and bound in Great Britain

Analytical Chemistry

This series of texts is a result of an initiative by the Committee of Heads of Polytechnic Chemistry Departments in the United Kingdom. A project team based at Thames Polytechnic using funds available from the Manpower Services Commission 'Open Tech' Project has organised and managed the development of the material suitable for use by 'Distance Learners'. The contents of the various units have been identified, planned and written almost exclusively by groups of polytechnic staff, who are both expert in the subject area and are currently teaching in analytical chemistry.

The texts are for those interested in the basics of analytical chemistry and instrumental techniques who wish to study in a more flexible way than traditional institute attendance or to augment such attendance. A series of these units may be used by those undertaking courses leading to BTEC (levels IV and V), Royal Society of Chemistry (Certificates of Applied Chemistry) or other qualifications. The level is thus that of Senior Technician.

It is emphasised however that whilst the theoretical aspects of analytical chemistry can be studied in this way there is no substitute for the laboratory to learn the associated practical skills. In the U.K. there are nominated Polytechnics, Colleges and other Institutions who offer tutorial and practical support to achieve the practical objectives identified within each text. It is expected that many institutions worldwide will also provide such support.

The project will continue at Thames Polytechnic to support these 'Open Learning Texts', to continually refresh and update the material and to extend its coverage.

Further information about nominated support centres, the material or open learning techniques may be obtained from the project office at Thames Polytechnic, ACOL, Wellington St., Woolwich, London, SE18 6PF.

Contents

Study Guide

This Unit has been written to provide a knowledge of the theory and practice of ultraviolet/visible spectrometry for both qualitative and quantitative chemical analysis.

Of all the areas of analytical chemistry it is fair to say that either directly or indirectly most laboratory workers encounter some form of ultraviolet/visible spectrometry at some stage. This may be in such simple things as checking the amount of an im purity in water using a colour comparator, or the more complex multi-sample absorption measurements using an automatic analyser in a clinical laboratory. This Unit seeks to enable the non-specialist to acquire sufficient knowledge about the scientific rules, techniques, procedures and equipment used in ultraviolet/visible spectrometry to appreciate its role and value as an analytical tool.

In order to achieve these objectives the Unit first deals with the simple concepts of the production of coloured substances in various reactions and the relationship of these colours with characteristic wavelengths of light in the visible region of the spectrum. It shows how colour is related to chemical structure and the manner in which various chemical groups can lead to changes in maximum absorption values. The Unit deals in some detail with the quantitative aspects of ultraviolet/visible absorption spectra and the laws governing the development and use of calibration curves for quantitative analysis. This aspect is well illustrated with fully worked examples showing how such curves may be plotted from experimental data.

As the value of the data from ultraviolet/visible spectrometry depends very much on the type of equipment used, the Unit also deals with the different types of instruments available. These range from the simple optical comparators through to highly sophisticated high resolution double beam spectrometers.

The Unit has been designed to provide a sound working idea of ultraviolet/visible spectrometry. But it does not cover the more obscure and abstruse mathematical concepts and calculations more generally associated with advanced physical chemistry. We are concerned here with the importance of ultraviolet/visible spectrometry in analysis for the measurement of real, everyday things encountered in many walks of life. This is applied chemistry for quality control, chemical purity, medical diagnosis and new product development.

Throughout the Unit there is a range of problems devised to enable you to gain a working knowledge of the procedures used to carry out quantitative determinations and to relate certain structures with calculated absorption wavelengths. But this is all really only a beginning and ideally the knowledge contained here needs reinforcing with practical experience of the methods and instrumentation wherever possible.

Bibliography

1. ANALYTICAL CHEMISTRY TEXTBOOKS

All books in this subject area contain major chapters dealing with ultraviolet/visible spectrometry. The following are used extensively:

(*a*) J. Bassett, R. C. Denney, G. M. Jeffery and J. Mendham, *Vogel's Textbook of Quantitative Inorganic Analysis*, 4th edn., Longmans, 1978.

(*b*) F. W. Fifield and D. Kealey, *Principles and Practice of Analytical Chemistry*, 2nd edn., International Textbook Company Ltd, 1983.

(*c*) H. H. Willard, L. L. Merritt, J. A. Dean and F. A. Settle, *Instrumental Methods of Analysis*, 6th edn., Wadsworth Publishing Co., 1981.

(*d*) D. C. Harris, *Quantitative Chemical Analysis,* W. H. Freeman & Co, 1982.

2. TEXTBOOKS ON ABSORPTION SPECTROMETRY

These books deal with the theory and practice of spectrometry with special chapters on ultraviolet/visible spectra treated in some depth.

(*a*) J. M. Hollas, *Modern Spectroscopy*, J. Wiley & Sons, 1987.

(*b*) C. N. Banwell, *Fundamentals of Molecular Spectroscopy*, 3rd edn., McGraw-Hill Book Co, 1983.

(*c*) D. L. Pavia, G. M. Lampman and G. S. Kriz, *Introduction to Spectroscopy*, Holt, Rinehart and Winston, 1979.

Practical Objectives

1. GENERAL CONSIDERATIONS

Even in a modestly equipped laboratory it is possible to carry out simple experiments to illustrate the basic concepts of ultraviolet/visible spectrometry. Many of the examples given in the early part of the text can be easily repeated in practice in order to gain experience. However, the application of your knowledge will be that much easier if you have access to a simple optical comparator or to a spectrometer.

2. AIMS

(a) To provide a basic experience of using

(i) a simple spectrometer or colorimeter for quantitative analysis.

(ii) a spectrometer for recording ultraviolet/visible spectra

(b) To give experience in the use of ultraviolet/visible spectrometry in quantitative analysis, with the emphasis on good analytical practice.

3. SUGGESTED EXPERIMENTS

(a) The operation of a recording and a non-recording ultraviolet/visible spectrometer to record spectra, determine λ_{max} values. The data to be used to calculate molar absorptivities and to plot Beer-Lambert Law calibration graphs.

(*b*) The determination of iron in potable waters.

(*c*) The enzymatic determination of glucose in a food stuff.

The three experiments suggested are closely linked to the content of this Unit. Other similar experiments could, however, be equally appropriate in realising the aims given above.

Acknowledgements

— Extracts in Section 3.1.1 on the determination of iron in raw and potable waters are taken from *Iron in Raw and Potable Waters by Spectrometry* (1977 version), published by H.M.S.O.

— The method reproduced in Section 3.1.2 for the determination of sucrose/glucose is reproduced by permission of Boehringer Mannheim GmbH.

— Extracts in Section 3.1.3 on the determination of trace amounts of iron are taken from the *Analyst*, **101**, 974–81, 1976 and reproduced by permission of the Royal Society of Chemistry.

1. Introduction

The foundations of quantitative chemical analysis can be traced back to the development of titrimetric analysis in which titration end-points depended on the change of colour of either the species being analysed or of that of a specially added chemical indicator. These colour transitions arise due to molecular and structural changes in the substances being examined, leading to corresponding changes in the ability to absorb light in the visible region of the electromagnetic spectrum. In various ways absorption spectroscopy in the visible region has long been an important tool to the analyst. Many important and sensitive colour tests have been developed for the detection and determination of a wide range of chemical species, both inorganic and organic in nature, and were first used long before the development of ultraviolet and visible spectrometers.

Today the uv/visible spectrometer is often referred to as the workhorse of the analytical laboratory, and is applied to many thousands of determinations which have been developed over the years. Uv/visible spectrometry has proved particularly useful in biochemical analysis, and is of vital importance in the clinical laboratory attached to most of our modern hospitals where various components of blood and/or urine in particular are determined and monitored on a 24-hour basis. It plays a part in environmental studies on pollutants, in forensic science work on drugs, and in maintaining the quality of the food we consume. In all of these realms analytical chemists and laboratory technicians regularly use uv/visible spectrometry as an essential tool in the identification and quantification of a very broad range of chemical and biological substances. The equipment for these purposes ranges from very simple colour comparators through to large computer controlled automatic scanning instruments covering the whole of the uv/visible region of the electromagnetic spectrum. But in all instances these studies involve measurement of radiation intensity at the spectral wavelengths which are characteristic for the substances under investigation.

1.1 COLOUR TESTS AND CHEMICAL ANALYSIS

One of the earlier tests you may have encountered in your study of chemistry is the change in colour of anhydrous copper sulphate crystals from white to blue when water is added, or the colour change of litmus paper from blue to red when the paper is dipped into a solution of an acid. These, and similar observations of colour and colour change, would probably have been your first experiences of the principles of absorption spectroscopy applied to the study of chemical systems. Thus a material will appear coloured if it shows selective absorption of radiation within the visible region of the electromagnetic spectrum and any change in that absorption will be associated with a change in colour.

The observation of colour or colour change has often captured the imagination of research workers and has led to significant discoveries in the field of chemistry and the development of new materials and dyestuffs. You should be able to think of a number of situations from your own experience where the generation of a coloured species or a sudden change of colour has caused you to observe the reaction system more carefully.

1.1.1 Colour Tests and Qualitative Chemical Analysis

For those responsible for the analysis of materials, simple colour tests have often provided useful preliminary or confirmatory evidence for the presence of particular chemical species. Thus in the early stages of your training in chemical analysis you will have already encountered a number of characteristic colour tests. Some possible examples are listed below:

(*a*) Use of litmus or indicator paper to test acidity/alkalinity.

(*b*) The yellow colour imparted to a gas flame when testing for sodium.

(*c*) The blue colour produced when testing for iodine with starch solution.

(*d*) The change in the colour of the crystals from yellow to green in the tube-and-bag breathalyser test, indicating the presence of ethanol.

No doubt some of these tests will be familiar to you. Six further well known tests are listed below.

Π Can you identify which of these tests involve an observation of colour or colour change? If you find this an easy question to answer then perhaps you could try to indicate the chemical reaction involved in each test!

(*a*) The test for halide ions using silver nitrate solution.

(*b*) The test for Mn^{2+} ions in aqueous solution with hydrogen sulphide gas.

(*c*) The test for Fe^{3+} ions with aqueous potassium thiocyanate solution.

(*d*) The test for aldehydes with Fehling's solution.

(*e*) The test for ketones with 2,4-dinitrophenylhydrazine.

(*f*) The Lassaigne test for nitrogen in organic compounds.

The correct response is that the appearance of a characteristic colour is the main observation in tests (*c*), (*d*) and (*f*). But with (*a*), (*b*) and (*e*) the main observation is the formation of a precipitate, although the colour of the precipitate may also be of some significance as indicated below. As for the nature of the chemical reaction involved this is indicated in the following notes.

Test a

The presence of a halide (X^-) is indicated by a precipitate of AgX which for chloride is normally white, for bromide is a very pale yellow colour and for iodide is distinctly yellow. Hence the colour of the precipitate is a useful part of the test.

$$X^- + AgNO_3 \rightarrow AgX + NO_3^-$$

Test b

A number of metal sulphides including copper, cadmium and lead are readily precipitated in acid solution, however manganese ions are precipitated under alkaline conditions, the precipitate being a pale creamy pink colour, which aids identification.

$$Mn^{2+} + H_2S \rightarrow MnS + 2H^+$$

Test c

Iron(III) gives a characteristic blood red colour with potassium thiocyanate, and this is a sensitive test for low concentrations of iron in the 3+ state. The formula of the complex ion can vary depending on the reaction conditions although the principal species will be the one shown.

$$Fe^{3+} + 3SCN^- \rightarrow Fe(SCN)_3$$

Test d

Fehling's solution is an alkaline solution of copper (II) complexed with tartrate ion which is initially blue in colour, but in the presence of aldehydes the blue colour is discharged and red copper(II) oxide is precipitated.

$$RCHO + Cu^{2+} \rightarrow Cu_2O + RCOOH$$

Test e

Both aldehydes and ketones react through the C=O (carbonyl) group with 2,4-dinitrophenylhydrazine to give solid precipitates of yellow or red hydrazones, possessing characteristic melting points.

$$R_2C{=}O + NO_2\text{-}C_6H_3(NO_2)\text{-}NHNH_2 \rightarrow R_2C{=}N\text{-}NH\text{-}C_6H_3(NO_2)\text{-}NO_2$$

Test f

The Lassaigne test for nitrogen involves the formation of sodium cyanide which when boiled with an iron(II) sulphate solution results in the characteristic Prussian blue colour of the product.

$$CN^- + Fe^{2+}/Fe^{3+} \rightarrow Fe_3(Fe(CN)_6)_2$$

If you gave the correct answer to four or more of the questions then you did very well.

Before looking at the nature of the light absorption which leads to colour, it is of interest to consider some examples from the early history of chemical analysis, in which the observation of colour played an important part.

1.1.2 Some Historical Aspects of Colour Tests

Many of the chemical reactions discovered in the late 18th century were utilised in the development of methods of analysis in the early 19th century and most of them involved either characteristic coloured reactions or precipitations of the type described above.

For example the preparation of Prussian Blue was discovered by a colour manufacturer in Berlin in 1704 when he boiled iron(II) sulphate with a solution of potassium carbonate contaminated with cyanide. Later a French dyer digested Prussian Blue with potassium hydroxide and obtained 'yellow prussiate of potash'. This was later called potassium ferrocyanide and now has the chemical name potassium hexacyanoferrate(II), $K_4Fe(CN)_6$. As there was much preoccupation with mineral analysis at the time, potassium ferrocyanide quickly became the preferred reagent for the identification of iron(III). W.T. Brande at the Royal Institution in London extended this study and found that the reagent was capable of giving characteristic coloured precipitates with sixteen metal ions including titanium and uranium!

However, it was not until 1822 that the corresponding reagent 'red prussiate of potash', potassium ferricyanide now known as potassium hexacyanoferrate(III), $K_3Fe(CN)_6$, was prepared by passing chlorine through a solution of the yellow prussiate. The ferricyanide was quickly adopted as the appropriate reagent for testing for iron(II).

Scheele was another researcher who encountered numerous colour changes of compounds during his work which led to the discovery of chlorine in 1774. He fused manganese dioxide (called black magnesia at the time) with potassium nitrate and extracted the solid mass with water to obtain a green solution which turned purplish-red on dilution. In the process, he converted potassium manganate(VI), K_2MnO_4, to the manganate(VII) or permanganate, $KMnO_4$. The colour change led Scheele to give the name 'mineral chameleon' to manganese ores. It was found that in acid solution the purple salt was easily reduced to the almost colourless Mn^{2+} species, a reaction which was soon applied in an increasing number of titrimetric analyses. Potassium permanganate solutions were first used in this way in 1846 to determine iron, and in 1858 for the determination of hydrogen peroxide and ethandioic (oxalic) acid.

These are just two examples of the importance of colour changes in the history of chemical analysis; and many others also exist. Since the ability to observe colour and to detect colour changes has been, and still is, of importance in analysis, it is relevant to consider the way in which light absorption produces the various colours and how we are able to interpret these absorptions as colours of different hues through our eyes.

1.1.3 The Electromagnetic Spectrum

One of the first statements made in the introduction to this part was that the appearance of colour arises from the property of the coloured material to absorb selectively within the visible region of the electromagnetic spectrum.

Are you aware how this region is defined and distinguished from other regions of the electromagnetic spectrum?

The visible region of the electromagnetic spectrum is defined in terms of the wavelength range to which the human eye responds. The usual range of wavelengths quoted is from 380 nanometres(nm) at the violet/blue end to 750 nanometres(nm) at the long wavelength red end of the spectrum. However, some sources quote wavelengths down to 360 nm as the short wavelength limit and up to 780 nm at the long wavelength limit, in recognition that the sensitivity of the human eye depends on the intensity of the radiation source, the conditions under which observation is made, and to some extent the individual making the observation.

You will probably be aware that beyond the short wavelength (violet-blue) end of the visible region there are (in order) the ultraviolet and X-ray regions, and beyond the long wavelength (red) and there are the infrared, microwave and radiowave regions. These regions are distinguished from the visible region by their characteristic wavelength values which vary from 0.1 nm for a typical X-ray wavelength to several hundred metres for radio waves. An alternative means of distinguishing these radiations is to quote their frequency values. Wavelength and frequency are related by the equation:

$$\text{Wavelength} \times \text{frequency} = \text{velocity}(c).$$

Since all electromagnetic radiation travels through space with a common velocity (the speed of light) $c = 3 \times 10^8$ m s^{-1}, it is therefore possible to calculate frequency values given the wavelength, and *vice-versa.* So that:

$$\text{Wavelength}(\lambda) = \frac{\text{speed of light}(c)}{\text{frequency }(\nu)} \quad (1.1)$$

Π Use the above expression and the given value of the velocity of light (c) to calculate the frequency of green light of wavelength 500 nm.

As a first step the 500 nm must be converted into 500 x 10^{-9}m. This ensures that identical units are used throughout the calculation. On dividing the speed of light by this value, you should obtain a value of 6×10^{14} s^{-1} (hertz), a value which shows that the corresponding frequency of a light wave is a very high value. The various regions of the electromagnetic spectrum are shown in Fig. 1.1a. It will be observed that the visible region is only a very small portion of an extensive range of radiation wavelengths.

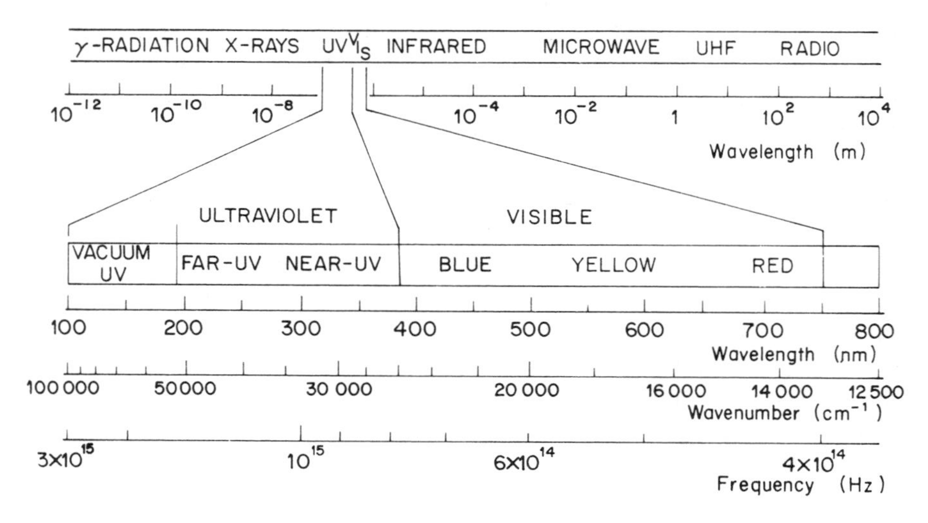

Fig. 1.1a. *The range of electromagnetic radiation*

What do we mean by the ultraviolet region, or to use a phrase common in uv spectrometry circles, 'the accessible ultraviolet'? The wavelength limits are commonly taken to be about 200 to 400 nm, although the lower limit is sometimes taken to be 190 nm and the upper limit 370 or 380 nm.

The lower limit is determined mainly by instrumental factors such as the lack of detector sensitivity, and reduced transmittance of radiation by optical components and by the oxygen in the air. An additional limitation is the marked reduction in the transmittance of radiation by common solvents at low wavelengths.

When investigating the wavelengths of uv/visible radiation at which a particular sample absorbs, it is useful to plot a spectrum which is simply a plot of absorption intensity versus wavelength. We will be meeting a lot of different spectra in this Unit, and the first one is given in Fig. 1.1b. Have a quick look at this figure, and note that the spectrum does not cover the entire region (200-800 nm) in which we are interested. The majority of spectra which you will come across will only cover a selected part of the uv/visible region, to reflect the particular wavelength range over which the particular sample absorbs.

In this particular Unit we are only concerned with the visible and the accessible part of the ultraviolet region covering the total wavelength range from about 200 nm to 800 nm.

SAQ 1.1a

Wavenumber ($\bar{\nu}$) values in cm^{-1} units are calculated by taking the reciprocal of the wavelength (λ) values, and multiplying by an appropriate factor to allow for the conversion of units.

(*i*) What is the relationship between wavelength values in nm and wavenumbers in cm^{-1} units?

(*ii*) Similarly what is the relationship between wavenumber values in cm^{-1} units and frequency values in units of hertz (or s^{-1})?

⟶

SAQ 1.1a (cont.)

Use these two relationships to calculate the wavenumber and the frequency of yellow radiation of wavelength 575 nm. Check your values against the wavenumber and frequency scales of Fig. 1.1a.

1.1.4 Light Absorption and Colour

Fig. 1.1b shows typical visible absorption spectra of a yellow, a red and a blue dye.

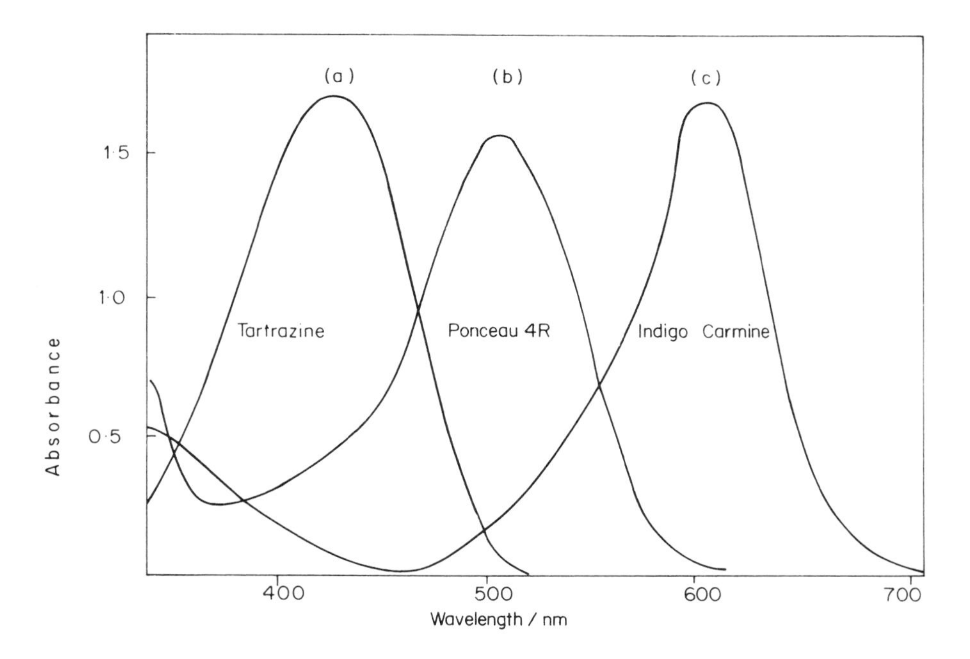

Fig. 1.1b. *Visible absorption spectra*

(*a*) Tartrazine or food yellow 4

(*b*) Ponceau 4R or food red 7

(*c*) Indigo Carmine or food blue 1.

These examples have been chosen to illustrate that the absorption band moves to longer wavelength as the principal colour changes in the order yellow, orange, red, purple, and blue. The colour observed is determined by the spectral distribution of the transmitted radiation, ie by the combined colour produced from the mixture of wavelengths *not* absorbed.

In order to define the absorption characteristics of a particular compound we could quote both the wavelength of maximum absorption, λ_{max}, and the range of wavelengths over which strong absorption occurs. In practice it is usual to quote the range of wavelength at half the absorption maximum (ie width at half peak height).

Π Can you obtain these two quantities (to the nearest 10 nm) for the three dyes shown in Fig. 1.1b?

You can compare your figures with those given below.

Colour of Dye (name)	Absorption Wavelength maximum/nm	Wavelength Range at half-height/nm
Yellow (Tartrazine)	430	380-480
Red (Ponceau 4R)	510	450-550
Blue (Indigo Carmine)	610	570-650

How did your results compare? Did you appreciate the meaning of the absorption range at half band-height?

Thus the yellow dye Tartrazine absorbs strongly over mainly the violet/blue region transmitting a mixture of green and red wavelengths which when seen together give the yellow colour observed. In contrast the bluish red dye Ponceau 4R absorbs in the blue to green region transmitting freely the red and violet wavelengths with this mixture again producing the observed colour. The general relationship between absorption position, colour of absorbed light and resulting colour observed for the transmitted light, is shown in Fig. 1.1c.

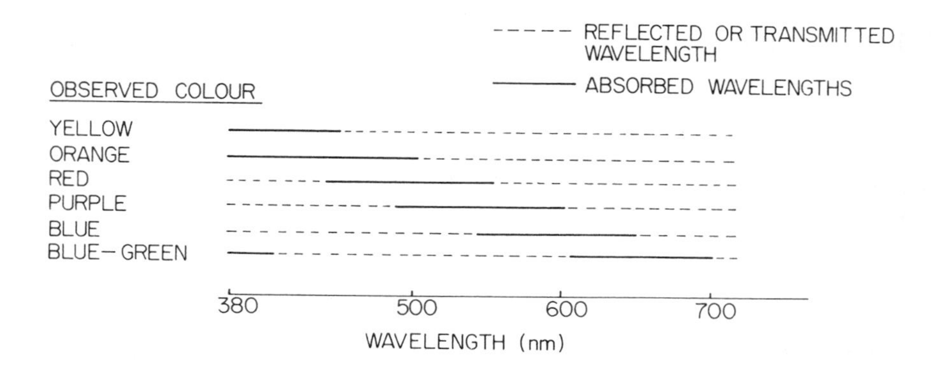

Fig. 1.1c. *Absorption band and colour relationships*

Note that the wavelength ranges listed define the positions at which the absorption maxima are observed. You may find that the limits and the ranges quoted in textbooks and other sources may differ slightly from those illustrated, although the trends will be similar.

One omission from the data in Fig. 1.1c is the wavelength range for a true (or single waveband) green to be observed. This is because to see a green colour it is necessary to have a material which absorbs at both the blue and red ends of the visible spectrum and transmits only in the middle or green region. This requires the species in solution to show two absorption bands in the visible region, or of course a green colour can be obtained by mixing yellow and blue dyes together in solution!

The above explanation is a simplification of the relationship between light absorption and observed colour. In particular the absorption spectra for the three dyes have half-band widths of 80 to 100 nm and are classed as narrow absorption bands, and thereby give rise to quite vivid colours. Most coloured materials, particularly solids, show much broader absorption bands than this. Also the fact that the human eye is reputed to be able to distinguish over 1 million different shades of colour illustrates the limitations of relationships such as those in Fig. 1.1c.

To test your understanding of the relationships involving position of absorption and colour you are asked to attempt the self-assessment question which follows. It involves distinguishing the absorption curves of some well-known reagents found in most laboratories and which can be readily identified by the colour of their solutions.

SAQ 1.1b

Using your knowledge of the colours of the common reagent solutions listed below, identify the solutions corresponding to the spectra A to E in Fig. 1.1d.

Solution	*Reagent and Concentration*
1.	aqueous copper sulphate solution (0.4 mol dm^{-3})

⟶

SAQ 1.1b (cont.)

Solution	*Reagent and Concentration*
2.	aqueous copper sulphate solution (0.04 mol dm^{-3})
3.	aqueous potassium dichromate solution (0.02 mol dm^{-3})
4.	potassium dichromate (100 mg dm^{-3}) in dilute sulphuric acid (0.005 mol dm^{-3})
5.	aqueous potassium permanganate solution (5×10^{-4} mol dm^{-3})

Comment, if possible, on the relative intensities of absorption (relative absorptivities) of the three compounds.

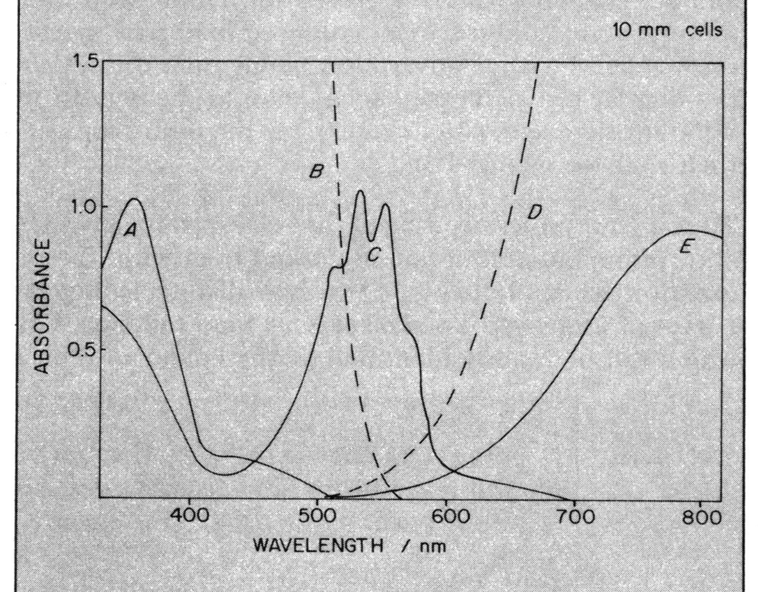

Fig. 1.1d. *uv/visible spectra of common reagents*

SAQ 1.1b

For a colour test to have high sensitivity it is important for the coloured material produced in the test to absorb strongly in the visible region. Thus as indicated in the response to SAQ 1.1b, the dichromate and copper sulphate solutions are weakly coloured because the main absorptions lie outside the visible region. To produce a sensitive colour test for chromium and copper species in aqueous solution, we need to use reagents which will produce species which absorb more strongly in the visible range.

For example a solution of 0.01 mol dm^{-3} $CuSO_4$ has a pale greenish blue colour, but when a few drops of concentrated ammonia solution are added, a whitish precipitate is initially obtained which redissolves to give a strong blue colour. The changes in the absorption spectrum are shown in Fig. 1.1e (note that these spectra extend beyond the end of the visible region).

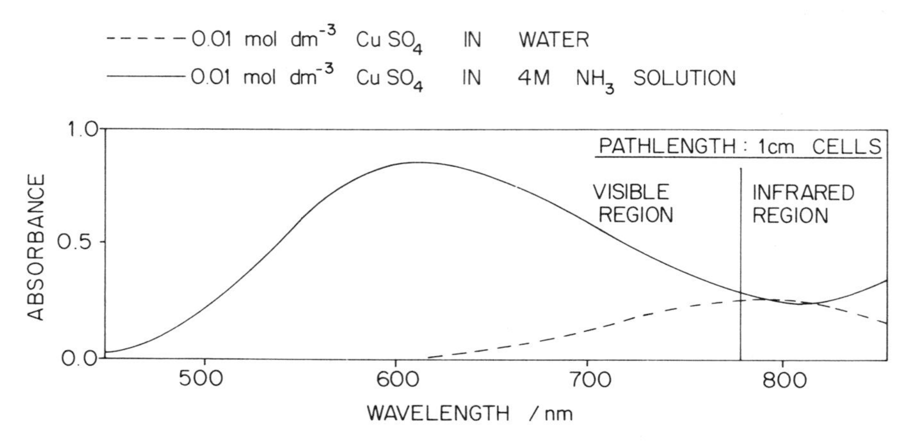

Fig. 1.1e. *Change in the absorption spectrum of copper sulphate upon addition of ammonia*

In this case the development of the stronger blue colour arises mainly because the absorption moves into the visible region, although there is also an increase in the intensity of the absorption. Some questions which might arise from these observations are:

(*a*) Is the formation of the blue colour characteristic of copper sulphate solution?

(*b*) What causes the increase in intensity of absorption when ammonia is added?

(*c*) Can the test be made quantitative?

(*d*) Why are both solutions essentially blue in colour, even though the absorption curves are significantly different?

The answers to the first three questions essentially form the subjects of Part 2 of the present Unit and the questions are covered again in SAQ 2.4b at the end of that Part.

1.2. THE BEER-LAMBERT LAW AND CALIBRATION DATA

Measurement of the absorption of ultraviolet and visible radiation by species in solution provides one of the most widely used methods of quantitative analysis available in the analytical laboratory. The bases of such measurements are:

(*i*) The generation of a suitably absorbing species in solution in amounts quantitatively related to the amount of analyte to be determined.

(*ii*) The selection of a suitable wavelength to enable accurate measurements to be made of the absorbing species.

(*iii*) The determination of the ratio of the intensity of the radiation on passing through a given thickness of the absorbing solution (usually held in a sample cell or cuvette of known path length) compared with the intensity of the same radiation beam when it passes through a reference cell containing the solvent or a suitable blank solution.

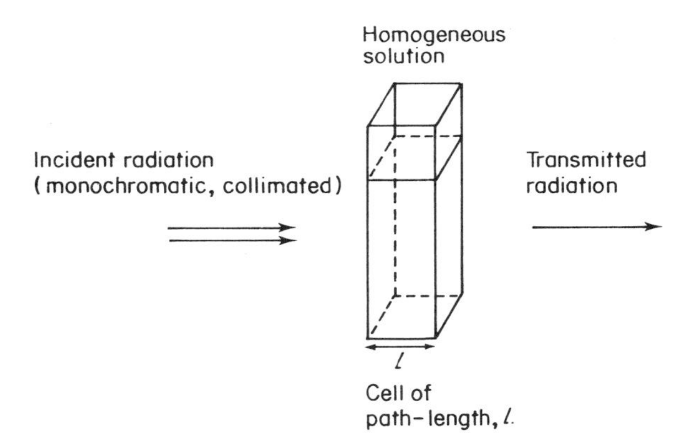

When a solution of the sample is contained in the cell the transmitted radiation has a certain intensity represented by the symbol I.

When the reference solution, normally the pure solvent, is contained in the cell, the transmitted radiation has an intensity represented by I_o.

The ratio of the radiation intensity transmitted by the sample to that transmitted by the reference solution is known as the transmittance, T, of the sample.

$$T = \frac{I}{I_o} \tag{1.2}$$

In practice we often measure percent transmittance, % T, because many instruments are calibrated with its very convenient scale of 0 to 100.

$$\% T = 100T = 100\frac{I}{I_o} \tag{1.3}$$

A more useful measure of the quantity of radiation absorbed is the *absorbance*. This is defined below.

1.2.1 Absorbance and Concentration

Suppose we have a solution of potassium permanganate which appears purple because it has a maximum absorption in the green at 530 nm, and we look at the way in which the intensity of green light changes as it passes through a 1 cm glass cell containing the solution. If we were able to show that after passing through 0.2 cm the light intensity had decreased to 50% of its initial value, then what is known as Lambert's law of light absorption would tell us that at 0.4 cm the intensity would have dropped by another 50% or to 25% of the original value. This would continue as shown in Fig. 1.2a, until by 1 cm the overall transmittance would be 3.125% of the original intensity. This plot of the transmittance against the path length is an exponential curve.

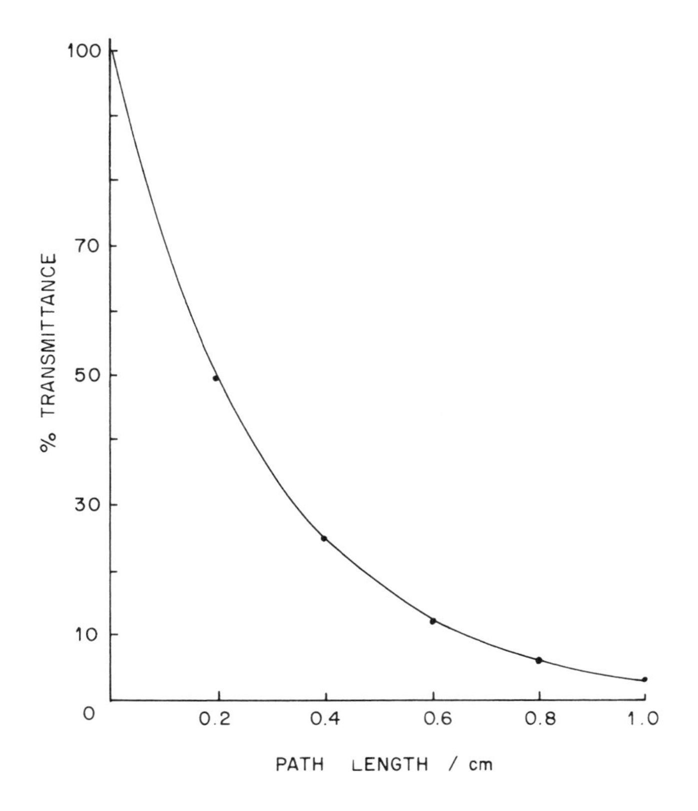

Fig. 1.2a. *Change of transmittance with path length*

To obtain a linear plot we need to use a logarithmic function of the transmittance,

and we now define absorbance (A) as

$$A = \log(I_o/I) = \log(1/T) \tag{1.4}$$

and

$$A = \log(100/\%T) \tag{1.5}$$

The table below shows the previous transmittance values converted into absorbances from which it will be seen in Fig. 1.2b that the plot of absorbance against path length is linear.

Path Length (cm)	0.0	0.2	0.4	0.6	0.8	1.0
% Transmittance	100	50.0	25.0	12.5	6.25	3.125
$T = \frac{I}{I_o}$	$\frac{100}{100}$	$\frac{50}{100}$	$\frac{25}{100}$	$\frac{12.5}{100}$	$\frac{6.25}{100}$	$\frac{3.125}{100}$
$\frac{1}{T}$	1	2	4	8	16	32
$A = \log\left(\frac{1}{T}\right)$	0.00	0.301	0.602	0.903	1.204	1.515

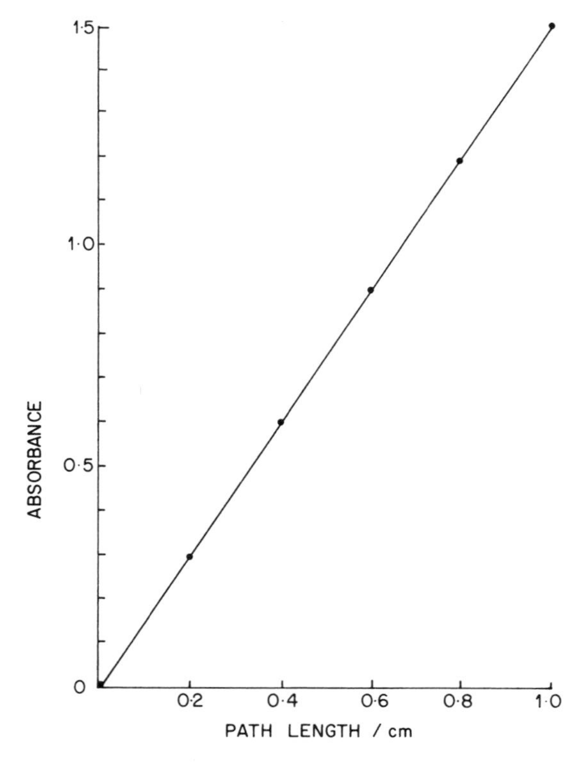

Fig. 1.2b. *Change of absorbance with path length*

The above treatment is for a fixed concentration of potassium permanganate, but another law, Beer's law, also tells us that the absorbance depends on the total amount of the absorbing species in the light path through the cell, and this means that the absorption is affected by both concentration (c) and path length (l).

Note that these two laws, first proposed in 1760 and 1852 respectively, apply strictly only to monochromatic light (that is light of a single wavelength) and depend on the absorbing system (the solution or solid through which the light is passing) being homogeneous.

The combined Beer-Lambert Law can be written in the simple form:

$$A = \text{acl} \tag{1.6}$$

where a is a proportionality constant known as the absorptivity. Absorptivity is a constant for a particular compound in a particular solvent at a particular wavelength, but it may have different numerical values depending on the units employed. We will return to this point shortly.

In order to apply this simple law to the determinations of an analyte species of unknown concentration in solution, it is necessary to first construct a calibration graph of absorbance *versus* concentration using standard solutions of known concentration of the analyte species. The absorbance of the 'unknown' can then be measured and the concentration interpolated from the calibration graph.

SAQ 1.2a

Carry out the following calculations:

(*i*) Obtain a value for the absorbance of a solution which only transmits 12% of the incident light.

(*ii*) Calculate the percentage of light transmitted for a solution with an absorbance value of 0.55. ⟶

SAQ 1.2a (cont.)

(*iii*) Determine the value for the absorbance of a solution of an organic dye (0.0007 mol dm^{-3}) in a cell with a 2 cm path length if its absorptivity is 650 dm^3 mol^{-1} cm^{-1}.

1.2.2 Preparation of Standard Solutions

The preparation of a suitable set of calibration solutions is common to most quantitative analytical procedures. But it is necessary for you to ensure that you can undertake the simple calculations involved in the preparation of such solutions and that you are aware of the criteria which determine the accuracy of such solutions.

For the analysis dealt with in the following section we require a series of solutions of potassium permanganate to contain manganese concentrations in the range 0 to 30 mg dm^{-3}.

To obtain the series of solutions to be measured, a stock solution was made by dissolving 0.0900 g of potassium permanganate in water and making the volume up to 250 cm^3 in a standard graduated flask.

The series of calibration solutions for spectrometric analysis of Mn were then prepared using the dilutions shown in the table below:

Solution	A	B	C	D	E
Volume of stock/cm^{-3}	2	4	6	8	12
Final diluted volume/cm^{-3}	50	50	50	50	50

Π Calculate the concentrations of the solutions A to E, expressing the results in:

(*a*) mg dm^{-3}, $KMnO_4$

(*b*) mg dm^{-3}, Mn.

(Relative atomic masses are K = 39.098, Mn = 54.938, O = 15.999). Comment on the factors which will determine the accuracy of the calibration solutions, mentioning the preferred titrimetric method of standardising permanganate solutions.

First we calculate the concentration, in mg dm^{-3}, of the stock solution

$$0.0900 \text{ g } KMnO_4 \text{ in } 250 cm^3 = 4 \times 0.0900 \text{ g dm}^{-3}$$
$$= 0.3600 \text{ g dm}^{-3}$$
$$= 360 \text{ mg dm}^{-3} \text{ of } KMnO_4$$

(*a*) Allowing for dilution:

$$2 cm^3 \text{ diluted to } 50 \text{ cm}^3 = 2/50 \times 360$$
$$= 14.4 \text{ mg dm}^{-3}$$

Hence tabulating the $KMnO_4$ content for the solutions:

Solution	A	B	C	D	E
Concentration/mg dm^{-3}	14.4	28.8	43.2	57.6	86.4

(*b*) To calculate the manganese content we use the ratio

$$Mn/KMnO_4 = 54.938/(39.098 + 54.938 + 63.996) = 0.3476$$

Hence 14.4 mg dm^{-3} $KMnO_4$ contain 0.3476 x 14.4 = 5.00 mg dm^{-3} Mn which gives

Solution	Blank	A	B	C	D	E
Manganese concentration/mg dm^{-3}	0.00	5.00	10.00	15.00	20.0	30.0

It must be remembered that the accuracy of the derived concentrations will depend on the accuracy of weighing and then, probably more importantly, on the accuracy of the volumes of solution measured and diluted.

Materials of the highest purity should always be used for the preparation of any standard solution reagent (Analar $KMnO_4$ is of 99.9% purity) and the material should be carefully dried before use. Sodium oxalate is recommended as a primary standard suitable for checking the strength of the potassium permanganate stock solution titrimetrically.

You may already be familiar with the preparation of standard solutions but revision of the simple calculation and some points about accuracy are often valuable. Notice that the blank solution in this case is a sample of the de-ionised water used to dissolve the potassium permanganate and to make the dilutions.

1.2.3 Calibration Data for Permanganate Analysis

Let us now look at some light absorption measurements on the solutions prepared in the previous section. Can you remember the first requirement for such an application?

The first requirement is to have an absorbing system that is stoichiometrically related to the analyte to be determined. In this case we have selected a manganese compound, $KMnO_4$ which has a stable (and characteristic) absorption in the visible region. Secondly we have to choose a suitable wavelength for measurement. This would normally be specified in the details of the analytical procedure being followed (see Part 3), or would have been determined previously and noted. In the present case we make no assumptions and record the spectrum over the visible region (do you remember the limits?), or if we did not have a recording instrument we would make measurements near the expected position of the λ max, and find which wavelength gave a maximum absorbance for one of the solutions.

We have in fact chosen to record the transmittance spectrum over the visible region for all the solutions A to E prepared in the previous section. This allows you to see that the shape of the transmission spectrum does not change much with concentration. All that occurs is that the transmittance values decrease as the solution colour becomes more intense, ie as the concentration of the $KMnO_4$ increases. It also gives us an opportunity to do some calculations converting transmittance to absorbance.

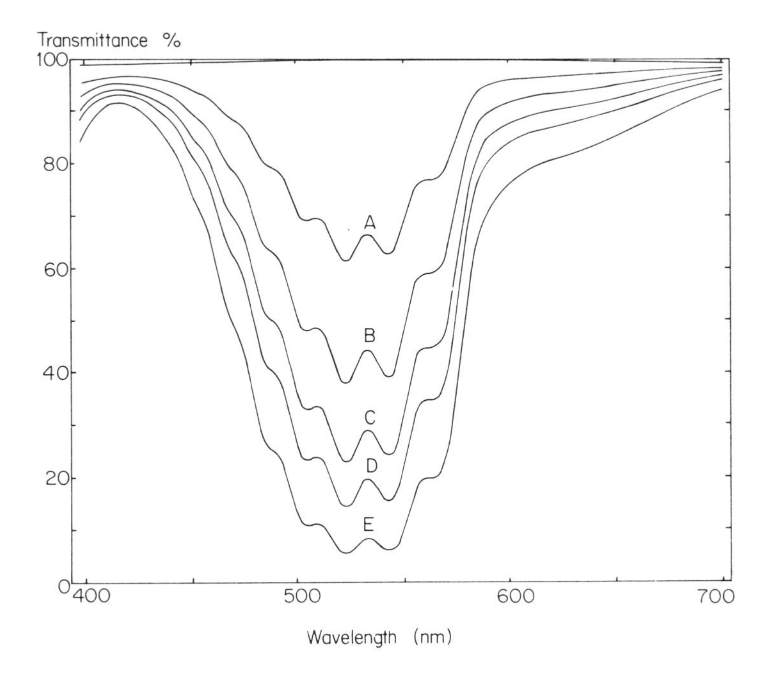

Fig. 1.2c. *Transmittance spectra of $KMnO_4$ solutions. The spectra from samples A to E represent increasing absorption (decreasing transmittance) in the green region of the spectrum*

For the solutions A to E the transmittance spectra in Fig. 1.2c were measured over the visible region (700-400 nm) with the permanganate solutions in 10 mm glass cells and using deionised water as the reference (or blank) cell.

You can now use these spectra to perform a simple exercise as follows:

Π (*i*) Read off, as accurately as you can, the wavelength of maximum absorption (λ_{max}) of $KMnO_4$ solutions.

(*ii*) Read off the transmittance values for each of the solutions at λ_{max} (noting that these are expressed as percentage values).

(*iii*) From these transmittance values calculate the corresponding absorbance values.

(*iv*) Plot your results as a calibration graph of absorbance against concentration of manganese.

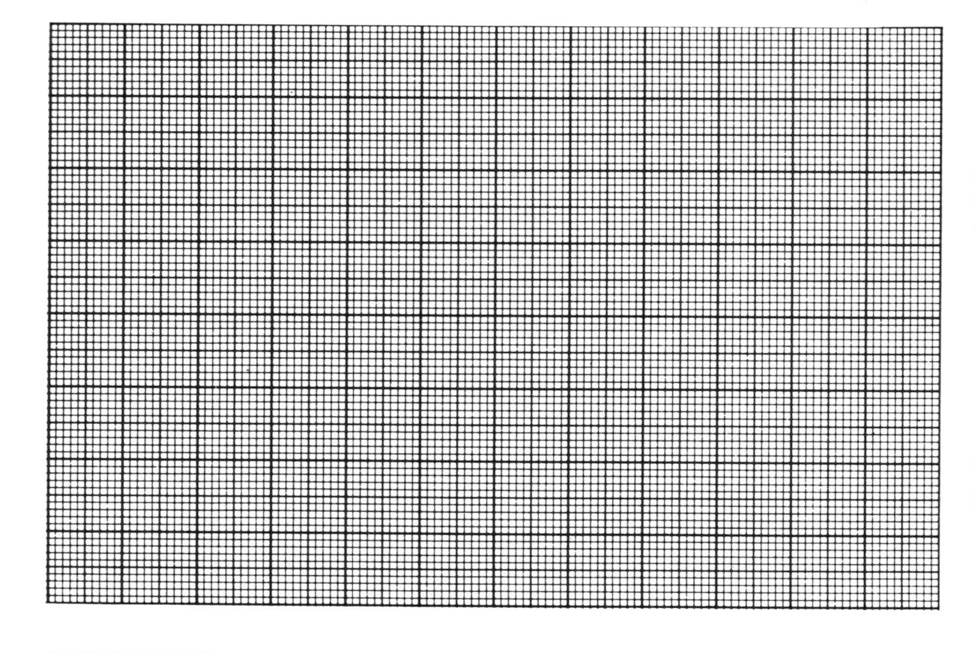

Did you get good agreement with the observed value of λ_{max} (522 nm)? Do your points lie close to the straight line, and does your calibration pass through the origin? These two last-mentioned criteria allow us to say whether the analyte obeys Beer's law over the concentration range tested. In this case adherence to Beer's law is good.

You should have the following set of results:

Solution	Blank	A	B	C	D	E
Mn concentration/mg dm^{-3}	0.0	5.0	10.0	15.0	20.0	30.0
% Transmittance at 522 nm	100	61	38	23	15	6
$\frac{I_0}{I}$	1.0	1.6	2.6	4.3	6.7	16
$A = \log \frac{I_0}{I}$	0.0	0.20	0.41	0.63	0.82	1.22

Before we leave this section let us examine the significance of absorptivity and calculate some values. You will recall that absorptivity is a constant for a particular compound under given conditions of solvent and wavelength. The numerical value, however, will vary with the units employed.

When we know the molar concentrations of the compound in question it is common practice to calculate the so called molar absorptivity (also known as molar absorption coefficient) and to give it a special symbol ε (epsilon).

Rearranging the Beer-Lambert Law, Eq. 1.6, and substituting ε for a we obtain.

$$\varepsilon = \frac{A}{cl} \qquad (1.7)$$

By using values of concentration and path-length in units of mol dm^{-3} and cm, respectively, we can calculate ε which will have units of $dm^3\ mol^{-1}\ cm^{-1}$. Note that absorbance is dimensionless and hence has no units.

Although the unit $dm^3\ mol^{-1}\ cm^{-1}$ has been widely used for many years it is now becoming more common to find ε quoted in the SI unit of $m^2\ mol^{-1}$. Let us calculate some absorptivities and then examine their significance.

The molar absorptivity for $KMnO_4$ can be calculated from the data given above. We will use the data for solution C:

$$KMnO_4 \text{ concentration} = 43.2 \text{ mg dm}^{-3}$$

$$\text{absorbance} = 0.638$$

$$\text{cell path length} = 1 \text{ cm}$$

$$\lambda_{max} = 523 \text{ nm}$$

The relative molecular mass of $KMnO_4$ is 158.032, and the molar

$$\text{concentration of solution C} = \frac{43.2 \times 120^{-3}}{158.032}$$

$$= 2.734 \times 10^{-4} \text{ mol dm}^{-3}$$

$$\varepsilon = A/cl$$

$$= 0.638/(2.734 \times 10^{-4} \times 1)$$

$$\varepsilon = 2334 \text{ dm}^3 \text{ mol}^{-1} \text{ cm}^{-1}$$

You will probably have realised that many analyses (and associated calculations of ε) are made at λ_{max}. This is because this gives rise to maximum sensitivity in the analysis. However, it is sometimes more convenient to work at other wavelengths and care must be taken to quote the actual wavelength chosen. When calculations of molar absorption coefficients are made at λ_{max} this is often indicated by adding the suffix to the symbol: ε_{max}.

In order to calculate the result in units of $m^2\ mol^{-1}$, we need to express concentrations in moles per cubic metre ($mol\ m^{-3}$) and path lengths in metres (m).

Applying this to our results above we have:

$c = 2.734 \times 10^{-1}\ mol\ m^{-3}$ (1000 times the previous numerical value),

and $l = 0.01$ m (1/100th of the previous value)

giving $= 0.638/(2.734 \times 10^{-1} \times 0.01) = 233\ m^2\ mol^{-1}$ (a value 1/10th of that previously).

Values of ε in SI units can therefore readily be obtained from published values in $dm^3\ mol^{-1}\ cm^{-1}$ by dividing the numerical quantity by 10.

Note that in quoting these results the figures have been rounded off to give the final answer to three significant figures. This is reasonable in the context of routine spectrometric analysis which involves manual operations such as dissolution and dilution, and reading results from a graph. Variations up to two to three percent are quite normal. Sometimes the relative molecular mass of the analyte species is unknown and it is impossible to calculate the molar absorption coefficient. In such cases it is usual to calculate

$$E_{1\%}^{1cm},$$

which is the absorptivity representing the absorbance of a 1% solution in a 1 cm path length cell.

What is the analytical significance of the absorptivity? When you have done the following calculations all will be revealed.

SAQ 1.2b

Calculate the concentration, in units of mg dm^{-3}, of a solution of each of the two compounds A and B. →

SAQ 1.2b (cont.)

Compound	M_r	$\varepsilon/\text{dm}^3\ \text{mol}^{-1}\ \text{cm}^{-1}$	Absorbance
A	250	1000	0.10
B	250	100 000	0.10

What is the significance of the molar absorptivity in analysis?

1.2.4 Visual Comparison

In all *visual comparison methods* we aim to match the intensity of the colour of two samples, so let us start by putting the transmittance of the test solution, 1, equal to that of the standard, 2.

Hence
$$T_1 = T_2 \qquad (1.8)$$

Because $\log (1/T_1) = \log (1/T_2)$ the absorbance of the two samples is also equal and

$$A_1 = A_2 \qquad (1.9)$$

If we are comparing identical compounds which obviously have identical spectral characteristics, then according to the Beer-Lambert Law

$$A_1 = \varepsilon_1 c_1 l_1 = \varepsilon_2 c_2 l_2 = A_2$$

and because $\varepsilon_1 = \varepsilon_2$

$$c_1 l_1 = c_2 l_2 \qquad (1.10)$$

this is the desired relationship for visual comparison techniques.

In the technique in which the dimensions of the two sample containers are identical ($l_1 = l_2$) and the samples are 'colour matched', then as you would expect $c_1 = c_2$.

In the alternative technique where colour matching is achieved by varying either l_1 and/or l_2, a simple calculation and instrument calibration is required, as we shall see.

Although visual colorimetry provides one of the simplest and cheapest methods of quantitative analysis it suffers from the following limitations:

(*a*) it is restricted to the analysis of coloured materials,

(*b*) it is sensitive to even low concentrations of coloured impurities,

(*c*) it suffers from a lack of precision.

For visual comparison techniques to be successful the colour of the standard and test solutions (as determined by the shapes of the visible absorption spectra) must be identical. Only the intensity of the colour ie the transmittance values should vary.

1.2.5 Types of Visual Comparators

The simplest visual comparison technique is to place a series of standard solutions of varying colour intensity in a set of test tubes of identical dimensions. The unknown test solution is then placed in an identical test tube and compared, in turn, with each of the standard solutions using the side-by-side viewing technique. The principal practical difficulty in using this technique is to ensure that the series of standard solutions covers the range of the test solutions with a sufficiently small increment in concentration to achieve the best realisable precision and accuracy. It is also necessary to prepare fresh standard solutions on a regular basis when determinations are to be carried out.

Use of a comparator such as the Lovibond Comparator eliminates the necessity of preparing standard solutions. With this equipment the set of standard solutions is replaced by a disc holding a series of permanent coloured glass filters designed to simulate the colour of the standard solutions. The discs are held in a special viewing box into which a tube containing the coloured test solution is placed alongside a tube containing only the colourless blank or solvent in line with the coloured filter.

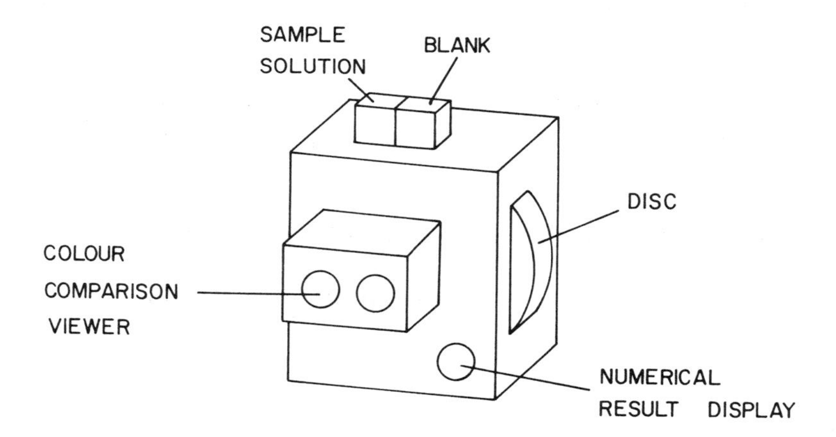

Fig. 1.2d. *Diagram of a disc comparator*

One of the main disadvantages of this technique is that the simulated colour provided by the disc is not always a complete spectral match to the test solution. This results in the colour match being sensitive to the background illumination used for viewing, and to a less extent on the colour vision characteristics of the analyst. With the Lovibond discs the recommended illuminant is average daylight.

The application of direct comparison methods is best understood from the use of the traditional Dubosq type colorimeter or visual comparator which avoids the requirement of a large series of standard solutions. This is achieved by allowing the depth of solution to be varied until a colour match is obtained with a single standard concentration. Thus Dubosq type comparators have viewing tubes which can be inserted into the unknown and a suitable standard solution and the relative lengths of solutions viewed are varied until a colour match is obtained.

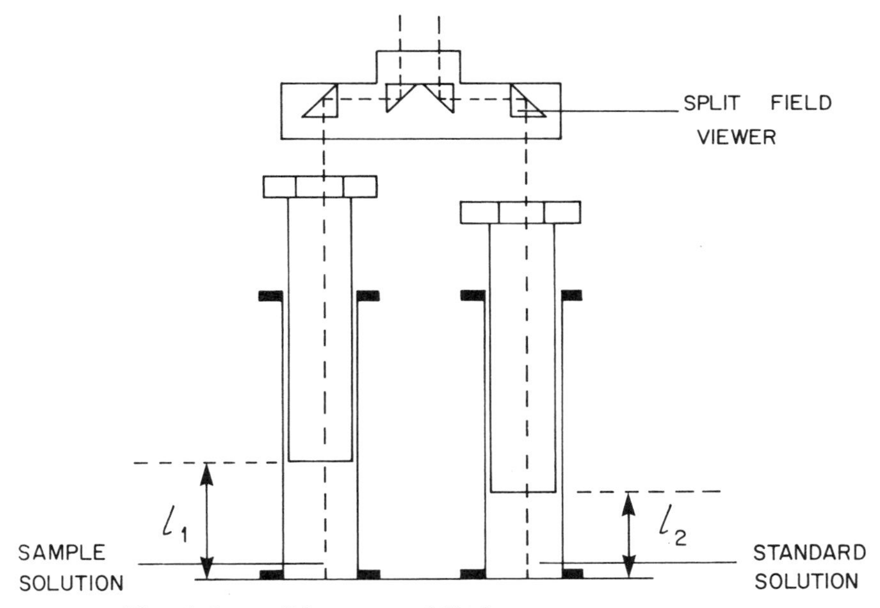

Fig. 1.2e. *Diagram of Dubosq type comparator*

The relative depths are indicated on calibrated wheels which control the depth of the viewing tubes. To improve the precision of the colour match point the light passing through the test and standard solutions are present in the eyepiece as two adjacent halves of the field of view. Thus if the standard of concentration c_1 requires a depth of l_1 to give the colour match to the test solution at depth l_2 then the required concentration of the test solution c_2 is calculated from:

$$c_2 = \frac{c_1 \, l_1}{l_2}$$

The following calculation illustrates the procedure with results from a Dubosq instrument.

∏ A Dubosq type comparator is being used to make measurements on permanganate solutions for the analysis of a steel for manganese content. (The manganese is oxidised in solution by bismuthate or periodate). A standard solution of 0.200 mg dm^{-3} Mn was set at a depth of 30.0 mm and the unknown was matched at a depth of 35.5 mm.

(*a*) Calculate the concentration of Mn in the unknown.

(*b*) If the ε_{max} of $KMnO_4$ is 2300 $dm^3\ mol^{-1}\ cm^{-1}$, use the Beer-Lambert law to calculate the absorbance and the % transmittance of the unknown solution at the path length quoted.

A_r (Mn) = 54.938

(*c*) Assuming a precision of ±3% in matching the transmittances, calculate the possible error in the unknown concentration.

(*a*) Applying the Dubosq relationship, $c_2 = c_1 l_1 / l_2$,

$$\text{Mn concentration of unknown} = 0.200(35.5/30.0)$$

$$= 0.237\ \text{mg dm}^{-3}$$

(*b*) Applying the Beer-Lambert Law, $A = \varepsilon c l$

Calculate the concentration in units of mol dm^{-3}

$$c = 0.237 \times 10^{-3}/54.938$$

$$= 4.23 \times 10^{-5}\ \text{mol dm}^{-3}$$

$$\text{Hence absorbance, } A = 2300 \times 4.23 \times 10^{-5} \times 3.55$$

$$A = 0.345$$

$$\text{since } A = \log(100/\%\text{T}) = 2 - \log\%\text{T}$$

$$\text{then } \log\%\text{T} = 2 - 0.345 = 1.655$$

$$\text{and } \%\text{T} = 45.2$$

(*c*) Applying a $\pm 3\%$ relative error to the transmittance gives

$$45.2 \pm (3 \times 45/100) = 45.2 \pm 1.4$$

The range of values is 43.8 to 46.6 %T

This corresponds to a range of absorbance values of 0.362 to 0.337, or 0.350 ± 0.013.

Since absorbance $\propto$ concentration, the concentration is

$$0.237 \pm 0.007$$

$$\text{or } 0.24 \pm 0.01 \text{ mg dm}^{-3}.$$

Merits of Visual Colorimetry

From the above discussion you should be able to list some of the advantages and disadvantages associated with visual colorimetry. How do you think the colorimetric determination of manganese in steel compares with say a volumetric method? Is the method quicker? Does it require a large amount of sample? Is the standardisation more difficult? These and many other similar questions must be asked before any judgement can be made on the relative merits of a particular type of analytical procedure. In the case of colorimetric methods the following advantages and disadvantages have been listed;

Advantages

(*a*) Colorimetric methods are usually rapid in comparison with volumetric and gravimetric methods.

(*b*) Often they require a minimum of sample preparation, sometimes only dissolution and colour development.

(*c*) Usually only a small amount of sample is required - in many methods a few milligrams will suffice.

(*d*) The simplest of equipment is required particularly with the visual comparison technique.

(*e*) Highly trained technicians are not required, as even non-technical personnel can be trained to do simple visual comparisons of colour intensity.

Disadvantages

(*a*) The preparation of fundamental standards for colorimetry can be a problem, and for visual comparison methods these may have to be replaced at frequent intervals.

(*b*) Where coloured filters are used to simulate the standards a suitable source of illumination must be specified and be available.

(*c*) The presence of interfering ions can cause colour distortions and invalidate the visual comparison.

(*d*) The sensitivity of visual methods is not high; an absolute accuracy of $\pm$ 5% may be expected routinely, which is much poorer than a good volumetric or gravimetric method of analysis.

1.3 BASICS OF UV/VISIBLE SPECTROMETERS

From what was said at the beginning of Section 1.2 about the criteria for applying light absorption measurements to the determination of an analyte in solution, can you list the requirements for a spectrometer to be used for such measurements? The main requirement is a knowledge of the range of wavelengths over which the analyte solution absorbs. If the solution is coloured then we immediately know that it absorbs over the visible range and hence an instrument operating over the visible region is probably sufficient. However, if you were expected to undertake biochemical analysis, an instrument capable of measuring in both the ultraviolet and the visible regions is likely to be required.

So the instrument must allow an appropriate wavelength to be selected suitable for the particular analyte. The sample and reference or blank solutions must be placed in the light beam in such a way that the ratio of the transmitted radiation beams can be measured. Finally the transmittance value or preferably the absorbance value for the solution should be displayed and recorded. These requirements allow us to list the basic components of a uv/visible spectrometer, and secondly to look at the way these components are assembled in a typical instrument.

1.3.1 Instrument Components

There are five essential components required for most absorption spectrometers. These are:

(*a*) A source or sources of radiation covering the required wavelength range.

(*b*) A means for selecting a narrow band of wavelengths - the device used for this is called the monochromator.

(*c*) Facilities for holding the cells or cuvettes containing the sample solution and the blank in the monochromated radiation beam.

(*d*) A device or devices capable of measuring the intensity of the radiation beam transmitted through the cells - this is the detector and is usually a photodetector.

(*e*) A display or output device to record the measured quantity in a suitable form.

The general arrangement of these components for a simple single beam spectrometer is shown in Fig. 1.3a

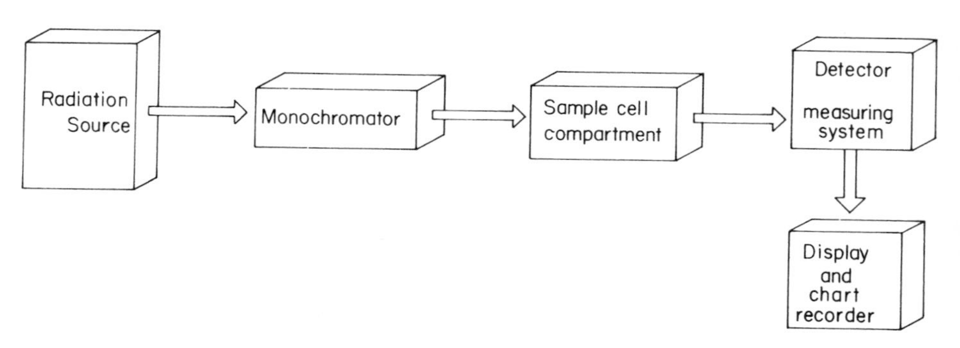

Fig. 1.3a. *Basic construction of a spectrometer*

The Source

Two sources are used in uv/visible spectrometers which between them can cover the whole range from 200 nm to 800 nm:

(*i*) For measurements above 320 nm compact tungsten-halogen sources in a quartz envelope are nowadays preferred as they give higher emission in the ultraviolet than the older tungsten filament lamps, which were restricted to measurements above 360 nm.

(*ii*) For measurements below 320 nm a deuterium arc source is used as this emits a continuous spectrum below 400 nm. Special filters are often included in the optical path when the tungsten-halogen lamp is being used below 400 nm. These are needed to reduce the chance of stray radiation reaching the detector and causing errors in the absorbance readings. (The effect of stray radiation is considered further in Section 2.4.3)

The Monochromator

The function of the monochromator is to select a narrow band of wavelengths to pass through the sample cell, and in most modern instruments is likely to be based on the use of a diffraction grating, arranged as illustrated in Fig. 1.3b.

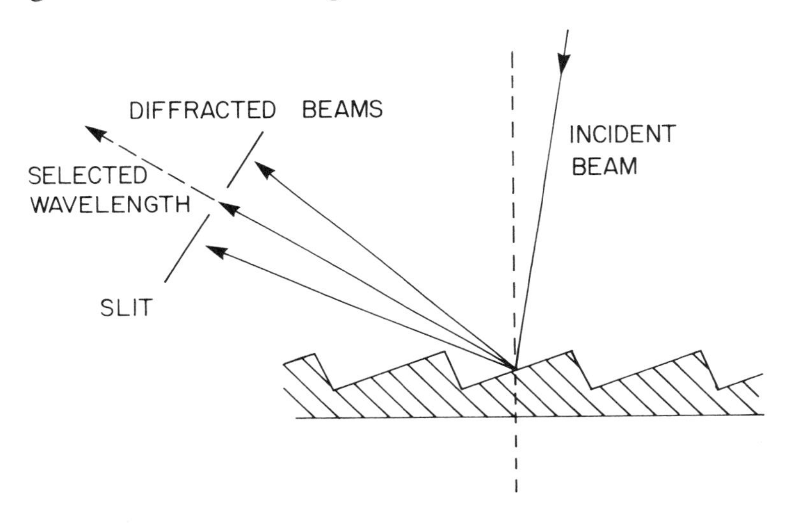

Fig. 1.3b. *Monochromation by a diffraction grating*

An important property of the monochromator is its bandpass (or bandwidth), ie the spread of wavelengths emerging from the exit slit for a given slit-width. The importance of bandwidth on measured absorbance is again dealt with in more detail in Section 2.4.2. One disadvantage of diffraction gratings is the possibility of different spectral orders emerging from the exit slit with a given angular and hence wavelength setting. Thus when a grating monochromator is set at 600 nm in its first order spectrum, the same angle will allow some 300 nm radiation through from the second order diffraction. A red filter is usually used in the light path when a grating monochromator is used above 600 nm to eliminate second and higher order transmissions. This problem did not occur with earlier instruments which used either narrow band filters or prisms for wavelength selection, but they lacked the quality of monochromation which is possible with diffraction gratings.

The Sample Cell or Cuvette

Cells are made of silica for the ultraviolet/visible, and glass or plastic for the visible. The optical windows are highly polished, flat and parallel and the light path between the inner surfaces of the windows is closely defined. The most commonly used cell is 10 mm in pathlength with a capacity of 3 to 4 cm^3 of solution. However, a wide variety of path lengths and cell volumes is available. For small volume cells correct horizontal and vertical location in the sample compartment is critical to ensure that the specified cell path length is used.

The Detector

The function of the detector is to respond to radiation falling on the sensing surfaces and to provide an electrical signal which is proportional to the intensity of that radiation. Two main types of detectors are currently used in uv/visible spectrometers. Silicon photodiodes are now gradually replacing the phototubes and photovoltaic cells incorporated in older instrument. Early silicon photodiodes had poor sensitivity below 400 nm but modern developments have improved their sensitivity so that they can be used to below 250 nm.

However for maximum sensitivity at low energies the photomultiplier tube is used in more expensive instruments. Photomultipliers have the advantage that they can be made to respond over the whole range from 190 nm to 950 nm. They need a high voltage supply connected to the various dynodes within the tube which are used to amplify the initial electron emission from the photocathode surface.

Output Devices

In single beam, manual spectrometers produced up to the early 1970's the output device was invariably a meter of some form which either indicated the transmittance directly or was used as a null point indicator in a potentiometric balancing circuit. The potentiometer control was usually calibrated both in transmittance units and (non-linearly) in absorbance units. Modern single beam instruments are more likely to have a digital output linked to a microprocessor so that the display gives absorbance values directly or can even be calibrated in appropriate concentration units after standard have been measured. The more expensive recording instruments give a graphical chart record of absorbance against wavelength, and in some cases the absorption spectrum will be displayed on a visual display unit along with appropriate numerical values.

1.3.2 Instrument Layout

The above section gave some brief details of the components which go to make up a uv/visible spectrometer. You are not expected to remember all the detail listed, but an understanding of the influence of the component characteristics on the reliability of the results produced should be the aim of all analysts. 'Get to know your instrument' is a good motto for the analyst who uses an instrument regularly. Don't treat it just as a black box, even although modern instrument design is leading us increasingly in that direction. It is always good to be able to at least read and understand the technical specification of an instrument.

For the present let us look at the optical layout of a typical single beam spectrometer, as shown in Fig. 1.3c.

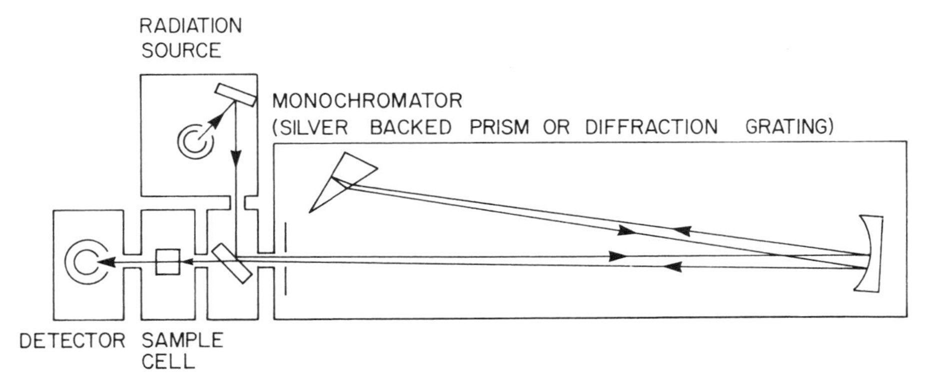

Fig. 1.3c. *Layout of a single beam uv/visible spectrometer*

The position of the main components will readily be observed, with additional front surfaced mirrors being used to direct the light beams from one component to another.

Can you indicate which of the five components is not shown in this figure? Yes it is the output device!

The arrangements in a recording instrument are similar except that the measuring beam is split into two components and passed through a sample cell and a reference cell simultaneously. This is usually done with a rotating sector mirror, so that the detector receives an alternating set of pulses, the intensity ratio of which give the necessary transmittance ratio measurements. The signals are usually separated and converted to read-out values electronically.

To test your general understanding of the basics of a spectrometer a self-assessment question follows which merely asks for the identification of the components as in the above figure, along with a question about the use of filters in the radiation path.

SAQ 1.3a

Fig. 1.3d below shows the optical components and layout of a typical recording spectrometer designed to operate over the wavelength range 190 to 900 nm. Identify on the diagram the four principal components:

Source

Monochromator

Sampling area

Detector

When the instrument is operated over certain wavelength ranges filters are inserted into the optical path. Specify which of the filters, red or blue, is used:

(*i*) at 780 nm

(*ii*) at 390 nm →

SAQ 1.3a (cont.)

In each case briefly explain the function of the filter used. What special precautions would you adopt to ensure the optimum instrument performance when measurements are being taken at 195 nm?

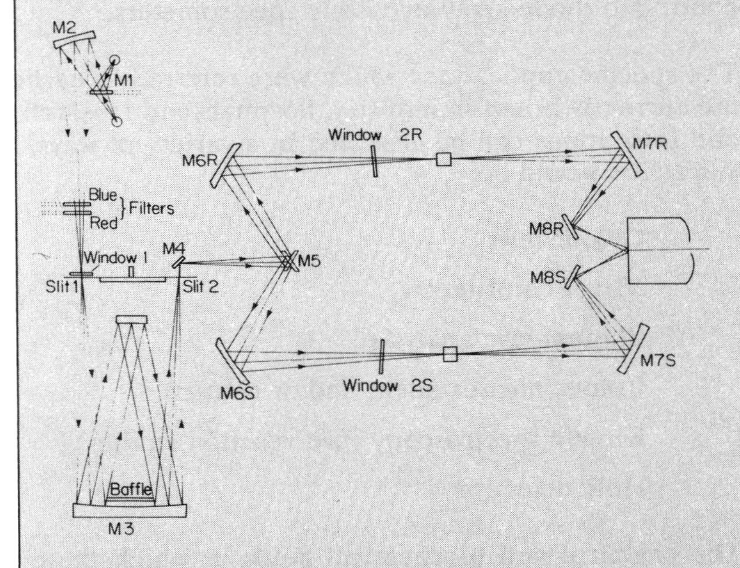

Fig. 1.3d. *Optical components and layout of a typical recording spectrometer*

1.4. SCOPE OF ANALYSIS BY UV/VISIBLE SPECTROMETRY

You will by this time be aware that measurement of the absorption of uv/visible radiation provides one of the most widely used means of analysis of chemical and biochemical systems. Included under this heading are techniques ranging from the simplest colour test to the most recently developed multi component analyses using computer-controlled diode array uv/visible spectrometers.

The specific applications which were referred to earlier and which are currently in use in industry, hospitals and research departments and institutions can be classified in a variety of ways. Typical classifications would be;

Colour tests

Visual colorimetry

Photometric analysis

Colour measurement and/or control

Kinetic spectroscopy (fast reaction studies)

Hplc detection

The chemical and biochemical fields in which these methods are applied can also be classified in a number of different ways eg,

Biochemical analysis

Enzymatic assays

Immunoassays

Pharmaceutical analysis

Trace metal detection

Vitamin analysis

Quantitative organic analysis

You may be aware of other methods of classification for a group of techniques based on uv/visible absorption. It will already be evident to you that these fields of application are so diverse that it is only possible in a Unit of the present size to emphasise the common principles involved in using uv/visible methods. However, it is important for you to appreciate the diversity of the general technique and in this Section an attempt will be made to illustrate that diversity in little more detail.

1.4.1 Colour Tests Today

The visual observation of a change in colour is so simple and straight forward that a number of regularly used analyses employ this principle. These include indicator papers for determining pH, (acidity and alkalinity) starch-iodide papers for detecting oxidising agents, heat test papers in the explosives industry, reagent papers and test sticks for medical diagnosis, and the latest 'dry-chemical' packs used for the clinical analysis of blood.

Such test papers or test kits can be designed to give both qualitative and quantitative information. Thus with the development of enzyme methods of analysis, and antigen-antibody complexes in the form of tests involving the observation of colour change, clinical biochemists are converting what were specialist analyses into forms suitable for use by doctors and nurses. The tests are designed to be robust, reproducible, highly specific and easy to carry out.

1.4.2 Visual Colorimetry (Using a Comparator)

The colour tests described above are a sub-set of the visual colorimetric methods of analysis, but in this section we look at some of the materials for which coloured glasses have been developed for use in the visual comparator of the Lovibond type mentioned in Section 1.2.5. The list below also includes materials for which the colour is used as an indication of product quality rather than being a measurement of chemical composition. Again the diversity of the substances should be noted.

Materials for which glass comparator colour standards are available include:

Air pollutants	Detergents	Paper
Alcohol	Disinfectants	Pharmaceuticals
Alloys	Drugs	Phenol
Aniline	Enamels	Plastics
Ashphalt	Fats	Resin
Beer	Fish	Rubber
Benzene	Flour	Rum
Biscuits	Foods	Sand
Bitumen	Glass	Sewage
Blood	Glucose	Silk
Butter	Glue	Soap
Caramel	Hair dyes	Spirits
Carotene	Honey	Sugars
Celluloid	Ink	Tobacco
Ceramics	Jams	Turpentine
Cheese	Jellies	Varnish
Chemicals	Lacquers	Vinegar
Cider	Lard	Water
Coffee	Leather	Wax
Condensed milk	Malt	Whisky
Cordials	Mercury	Wines
Cosmetics	Metals	Wool
Cresols	Milk	Yeast
Custard	Oats	
Dental materials	Paint	

1.4.3 Biochemical Spectroscopy

Uv/visible spectroscopy has played an important role in the study of natural products from plant and animal sources. These investigations have involved the steps of isolation, characterisation, synthesis and biosynthesis and then to biological modes of action; uv/visible spectroscopy has been of value in each of these stages for a wide range of biochemical materials. In the two volume publication 'Biochemical Spectroscopy' by R.A. Morton, some of the chapter titles give an indication of the range of natural products for which uv/visible spectral studies have been undertaken;

Carotenoids and related substances
Aminoacids, proteins and enzymes
Heterocyclic compounds including nucleotides and nucleic acids
Porphyrins, bile pigments and cytochromes
Steroids and related substances
Vitamins and coenzymes
Some antibiotics and medicinal substances

For instance, much of the work carried out in the USA on steroids by Professor Woodward involved using uv/visible spectrometry as a basis for assigning chemical structures to compounds often differing only by the location of a carbon-carbon double bond or of a carbonyl group. He was able to predict the wavelengths at which maximum absorptions would occur, simply from a knowledge of the component chemical groups in the molecules.

Although many of these natural products have characteristic uv/visible spectra when pure, it is not always easy to use spectrometric methods of analysis when they are present in complex mixtures. However, the analysis of specific components in complex mixtures such as those of the food industry can be achieved by enzymatic assays. Such assays are often based on the measurement of an increase or decrease in the absorbance of NAD (nicotinamide-adenine dinucleotide) at 340 nm (uv method) or NADH its reduced form, can be oxidised in the presence of a tetrazolium salt to yield a coloured dye (colorimetric method). Enzyme assays of this type have been developed for such substances as amino acids, carbohydrates and organic acids.

A typical procedure for the determination of sucrose and glucose in foodstuffs is given in Part 3 of this Unit and includes the enzymatic hydrolysis of sucrose to glucose and fructose prior to determination of the glucose by measuring the quantitative amount of the reduced nicotinamide-adenine dinucleotide phosphate (NADPH) produced in the process.

Direct photometric analysis of biochemical materials is now less frequent as it has been replaced by these enzymatic methods and more recently by immunoassay methods. However, complexometric methods using colour forming reagents are still used for determining the common metal ions such as calcium, magnesium and iron, as will be discussed in Part 2 of the Unit.

1.4.4 Uv/visible Monitors

Special small volume flow-through uv sample cells have been developed for high performance liquid chromatography (hplc) systems. This enables continuous monitoring of eluted components from the chromatographic column. The detection of individual species depends on their abilities to absorb the radiation used in the detector. When analyte species do not absorb radiation at the operating wavelengths of the detector it may be possible to detect them by incorporating an absorbing centre in the molecule by an appropriate chemical reaction. Such procedures are termed 'chemical derivatisation' or labelling reactions. A common example is the derivatisation of amino acids using dansyl chloride, as follows:

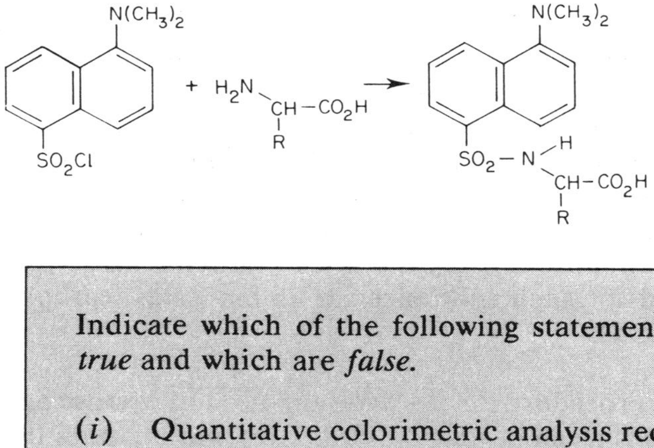

$N(CH_3)_2$
SO_2Cl
$+ \; H_2N-CH(R)-CO_2H \longrightarrow$
$N(CH_3)_2$
$SO_2-N(H)-CH(R)-CO_2H$

SAQ 1.4a

Indicate which of the following statements are *true* and which are *false*.

(*i*) Quantitative colorimetric analysis requires the use of a light measuring instrument.

(*ii*) The eye is as good as an instrument for detecting colour changes.

(*iii*) All colorimetric methods of analysis have been developed for trace levels. ⟶

SAQ 1.4a (cont.)

(*iv*) Certain biochemical methods of analyses can be designed to be highly selective, even though they all involve measurement of change in the uv spectrum of NAD (nicotinamide-adenine dinucleotide).

(*v*) The use of uv/visible spectrometry in the monitoring of the separation of mixtures by high performance liquid chromatography (hplc) is limited to those components which show strong absorption above 220 nm.

1.5 SPECTRA-STRUCTURE RELATIONSHIPS

So far we have dealt with the nature and measurement of uv/visible spectra but have not considered in any depth how these spectra arise. What structural features of molecules give rise to absorption of radiation at various wavelengths and hence lead to different colours being observed for different compounds?

This absorption occurs due to the fact that all molecules possess electrons which can be excited (raised to a higher energy level). Many of these electrons are excited by radiation of uv or visible wavelengths, but others are only excited by radiation within the vacuum ultraviolet region.

1.5.1 Electronic Spectra

If only transitions in electron energy levels were involved the uv/visible spectra for all compounds would consist of fairly sharp lines, ie very narrow absorption bands.

The energy difference between the electronic energy levels is given by the equation:

$$\Delta E = h\nu$$

In its simplest form this can be represented diagrammatically as shown in Fig. 1.5a in which the absorption of energy leads to electrons, initially in the ground state, moving to an excited state (an energy level of greater energy than the ground state).

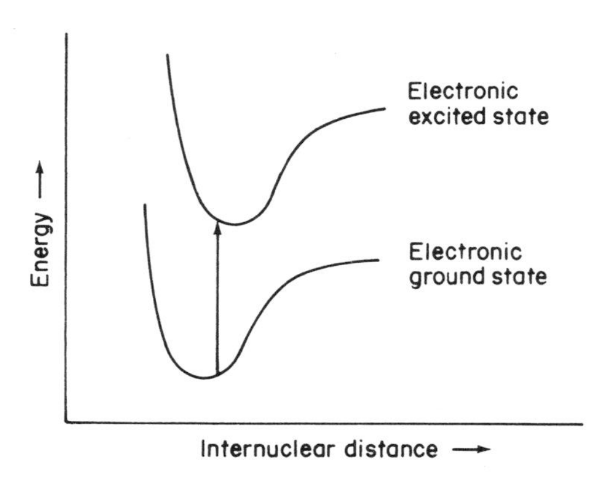

Fig. 1.5a. *Electronic transition energy diagram*

But as you have seen, most of the spectra are actually very broad smooth curves, and not sharp peaks. This is because any change in the electronic energy is accompanied by a corresponding change in the vibrational and rotational energy levels. You may already know that vibrational and rotational energy changes by themselves give rise to infrared absorption spectra. However, when they accompany ultraviolet/visible absorptions a large number of possibilities exist within each electronic state and the individual absorption bands normally become very broad.

The energy transition illustrated in Fig. 1.5a does not represent the full story as the change will actually be from a vibrational energy level in the electronic ground state to one of several vibrational levels within the excited state. You can see this more clearly in Fig. 1.5b below.

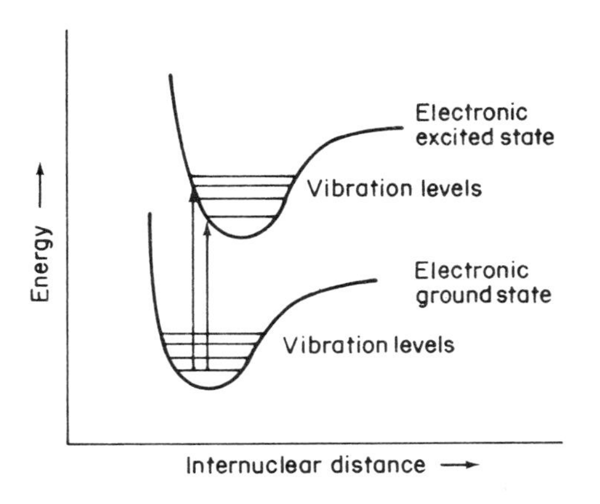

Fig. 1.5b. *Electronic transitions between vibrational energy levels*

In some instances the fine structure produced by the vibrational levels may be observed in the ultraviolet/visible spectra of compounds such as benzene and toluene in the gaseous state. But this is not normally observed with solutions and substances in the liquid state, although benzene does still produce a spectrum with a certain number of sharp well-defined fine structure peaks.

1.5.2 Structure and Energy

A variety of energy absorptions is possible depending upon the nature of the bonds within a molecule. For instance, electrons in organic molecules may be in strong σ bonds, in weaker π bonds or non-bonding (n). When energy is absorbed all of these types of electrons can be elevated to excited antibonding states which can be represented diagrammatically as in Fig. 1.5c, the antibonding states being represented with an asterisk as σ^* and π^*.

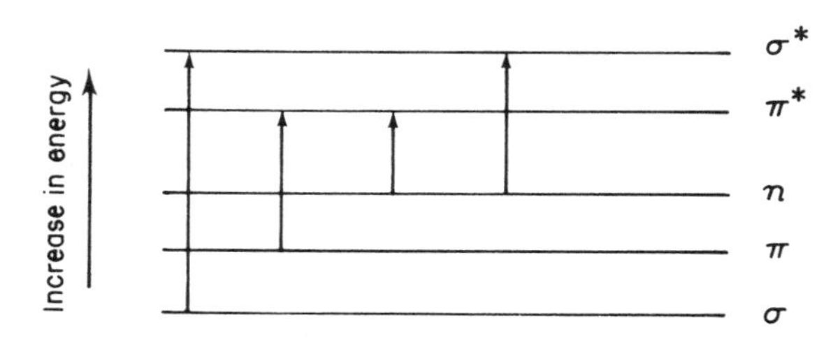

Fig. 1.5c. *Bonding and antibonding energy transitions*

Most σ to σ^* absorptions for individual bonds take place below 200 nm in the vacuum ultraviolet region and compounds containing just σ bonds are transparent in the near ultraviolet/visible region. $\pi \rightarrow \pi^*$ and $n \rightarrow \pi^*$ absorptions occur in the near ultraviolet/visible region, and result from the presence in molecules of unsaturated groups known as *chromophores.* These are dealt with more fully in Part 4, but at this stage you should know that chromophores have characteristic molar absorptivities and absorb at fairly well defined wavelengths. Some typical chromophores are listed in Fig. 1.5d.

Chromophore	Typical Compound	Electronic Transition	Characteristic values	
			λ_{max}/nm	ε/m^2 mol^{-1}
$>C=C<$	ethylene	$\pi \rightarrow \pi^*$	180	1300
$>C=O$	acetone	$\pi \rightarrow \pi^*$	185	95
		$n \rightarrow \pi^*$	277	2
	benzene	$\pi \rightarrow \pi^*$	200	800
			255	22
-N=N-	azomethane	$n \rightarrow \pi^*$	347	1
-N=O	nitrosobutane	$n \rightarrow \pi^*$	665	2

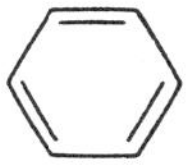

Fig. 1.5d. *Typical chromophores*

The wavelengths of these characteristic absorptions and their molar absorptivities are often greatly changed due to the presence of other chemical groups in the molecular structure. It is found that groups such as -OH, $-NH_2$ and halogens, which all possess unshared electrons, cause the normal chromophoric absorptions to occur at longer wavelength (ie displaced towards the red end of the spectrum) and with an increase in the value of the molar absorptivity. Groups which cause this change are known as *auxochromes.*

This means that compounds possessing several chromophores and auxochromes are likely to be coloured. You will also find that the greater the degree of conjugation in the molecule (by this we mean alternate double and single bonds) the longer the wavelength at which the ultraviolet/visible absorption will occur. Substances such as carotenes which contain eleven alternate double and single bonds appear bright red in colour because the extensive conjugation causes them to absorb strongly in the blue part of the visible region of the spectrum.

1.5.3 Absorptions in Inorganic and Organo-metallic Compounds

You will be well aware that many inorganic compounds, such as potassium permanganate and sodium dichromate are very highly coloured, although they do not possess the unsaturated, conjugated structures characteristic of organic compounds. The colour is due to transitions occurring in the energy levels of the d electrons in the transition metals, ie in the manganese atom of the potassium permanganate and the chromium atom of the sodium dichromate.

Where transition metals are also linked to organic ligands in organo-metallic compounds very intense colours can arise due to d $\rightarrow$ d, n $\rightarrow$ π^* and π $\rightarrow$ π^* transitions all taking place. Another type of transition which affects the colour of these compounds is known as charge-transfer, in which an electron occupying a σ or π orbital in the ligand is transferred to an unfilled orbital of the metal, and *vice-versa.* The strong colours resulting from these various effects means that the formation of organo-metallic compounds is a useful procedure for trace metal analysis as even very small quantities of the metals can be made to produce compounds with high molar absorptivities.

You will find the subject of structure, conjugation and colour dealt with in more detail in Part 4.

Summary

For many years the measurement of light absorption and transmission has served as a basis for the measurement of concentrations of substances in solution. As a result of this a series of mathematical equations have been developed relating the degree of light absorption with the effective solution path length, the nature of the substance and its concentration. Special instruments have been manufactured which enable the ultraviolet and visible radiation to be measured with a high degree of accuracy at different wavelengths.

Objectives

You should now:

- understand the importance of colour measurement in chemical analysis;
- be able to carry out calculations based upon the features of the Beer-Lambert Law;
- have acquired a knowledge of the functions of the components of a spectrometer;
- appreciate the importance and scope of uv/visible spectrometry in a wide range of applications.

2. Quantitative Methodology

Probably before embarking on this course you will have used a spectrometer to carry out at least one quantitative analysis on a coloured solution. Let us suppose that you are familiar with the method for determining the iron content of water, based on the formation of the red-orange complex of iron(II) with 1,10-phenanthroline (the most often used complexing agent for iron when present at low concentrations). The method of analysis to be used would be given in the standard methods for such analysis used in the water industry, or suitably adapted therefrom.

Π Can you suggest what details would be specified in a typical procedure involving a spectrometric method of analysis of an inorganic component in a sample of tap water?

You can check your answer against the following list:

(*a*) the amount of material to be used,

(*b*) the amounts of reducing agent, complexing agent (the 1,10-phenanthroline) and buffer solution to be added to give the required orange-red coloured complex, and the time to wait before the solution is ready for measurement,

(*c*) the preparation of a set of calibration solutions,

(*d*) the wavelength selected for the measurement, and the cell dimensions to be used.

(*e*) the method of calculating the analytical results from the experimental measurements.

How many of these points did you list – at least four? – good; all of them plus some others! – excellent; you obviously have a good grasp of the principles of a spectrometric analysis.

Π What other steps would have to be specified for the determination of an inorganic component in an organic matrix such as a food product, or in a water sample with a high organic content?

The organic matrix may well have to be removed because it is very likely to cause severe interference problems. Two additional steps are:

(*a*) Removal of the organic matter, either by wet oxidation or by ashing

(*b*) Solution of the residue by an acid dissolution procedure.

Some examples of the detailed specifications of spectrometric analysis are given in Part 3 of this Unit, but one object in the present Part is to consider some of the factors which are important whenever a spectrometric analysis in the uv/visible region is attempted.

Returning to the determination of the iron content of water, some questions that might be asked include:

Is 1,10-phenanthroline the best reagent to use for iron?

Why is a reducing agent needed?

Why do we add the buffer solution?

If time is important does that mean that the colour of the complex is unstable?

Do we always make the measurement at the wavelength of maximum absorption?

Can we use plastic cells to hold the solution to be measured?

These and other questions are usually considered by scientists who test proposed analytical methods before they are adopted as standard methods, as for example in the water industry.

Thus, when a sample is to be analysed by uv/visible spectrometry, the method of sample preparation, solution conditions of measurement and instrumental parameters to be used will all have been carefully considered and standardised during the design of the analytical procedure.

In the ideal situation the desired analyte species would be easily isolated, and converted to a highly absorbing form possessing a characteristic absorption band within the range of the instrument available. The absorbing species should be stable and unaffected by solution conditions. Interference from other components in the solution should also not occur. The system should also obey Beer's law over a wide concentration range, and the linear calibration graph should be reproducible and insensitive to small changes in instrumental characteristics.

However, ideal conditions rarely exist in practice, and the purpose of this Part is to examine the criteria which enable the optimum conditions to be selected for reliable spectrometric analysis.

2.1 SOLUTION PREPARATION, SOLVENTS AND CELLS

The appropriate treatment for obtaining a solution of the desired component (the analyte) is determined by the nature of the sample, the component to be determined, the other constituents present, the desired accuracy and the time available. We will therefore assume that we have a solution of the absorbing constituent, the absorption either being a property of the original analyte or of a chemical derivative of that analyte.

2.1.1 Stability and Solubility

To be suitable for spectrometric measurement in the uv/visible the analyte solution should possess the following properties.

(*a*) The absorbing species should have stability for sufficient time to permit accurate measurement to be made. Obviously, if the absorption is changing as you attempt to make the measurement then the result will have poor accuracy. Instability can arise as a result of many factors, such as air oxidation, photochemical decomposition or due to solution conditions such as solvent, pH, temperature. In some cases little can be done to improve stability and it is then very important that the procedures used and the interval of time before measurement are carefully standardised, so that any deterioration is constant.

(*b*) Colloidal or insoluble material must not develop due to slow hydrolysis or some other type of reaction with the solvent.

Reactions yielding a colloidal system or suspension are difficult to control and are generally susceptible to the presence of electrolytes and other constituents. Stabilising additives such as agar and gelatin are only partially satisfactory. If the product is insoluble it can sometimes be extracted into another solvent.

The principal effect of the presence of colloidal or suspended material is an apparent increase in absorbance due to light scattering and, since this is significantly wavelength dependant, the effect is much more noticeable in the ultraviolet (say below 300 nm) than in the long wavelength visible region (say above 500 nm).

2.1.2 Choice of Solvent

From the above discussion you should be able to deduce that the important characteristics of the solvents are;

(*a*) good solubilising power

(*b*) stable interactions with the absorbing species

From the optical point of view the solvent should be transparent in the region of measurement, and should be of consistent purity.

Water is the cheapest and most transparent solvent and is commonly used for water soluble susbtances. Unfortunately only a small proportion of organic compounds are water soluble and organic solvents are needed. A practical difficulty in the use of water in closed cells is its tendancy to give rise to air bubbles which cause errors due to scattering the light passing through the cell. The best remedy is to use freshly boiled water which contains little or no dissolved air. Water which has been distilled or deionised is generally of sufficient purity, but water which has been stored in plastic containers may contain small amounts of uv absorbing impurities.

The alcohols (methanol, ethanol and propan-2-ol) also have good solubilising power and can have good transparency in the low uv region when carefully purified. Hexane and cyclohexane are also transparent in the uv region provided any traces of benzene or other aromatics have been removed before use (by passing through a silica column or shaking with sulphuric acid; alternatively, use Spectroscopic Grade solvents).

Solvents such as trichloromethane, tetrachloromethane and the ethers (diethyl ether, diethyl cellosolve and dibutyl carbitol) are commonly used in solvent extraction procedures, particularly of metal complexes, but have limited transparency in the uv below 300 nm. Diethyl ether is also unsuitable because of its high volatility.

The effective cut-off wavelengths in the ultraviolet region for a range of solvents is given in Fig. 2.1a.

Solvent	λ(nm)
Hexane	199
Heptane	200
Isoctane	202
Diethyl ether	205
Ethanol	207
Propan-2-ol	209
Methanol	210
Cyclohexane	212
Acetonitrile	213
Dioxan	216
Dichloromethane	233
Tetrahydrofuran	238
Trichloromethane	247
Tetrachloromethane	257
Dimethyl sulphoxide	270
Dimethyl formamide	271
Benzene	280
Pyridine	306
Propanone	331

Fig. 2.1a. *Cut-off wavelengths for common solvents. Values at which the transmittance falls to 25 per cent ($A = 0.602$) measured in 10 mm cell using water as the reference*

2.1.3 Sample Cells (Cuvettes)

Cells for the visible region may be made of glass (or transparent plastic if only aqueous solutions are being measured), but for the uv region below 330 nm quartz or fused silica cells are necessary. A cell is specified by its type, material of construction, pathlength and dimensional tolerance.

Some typical cells are shown in Fig. 2.1b.

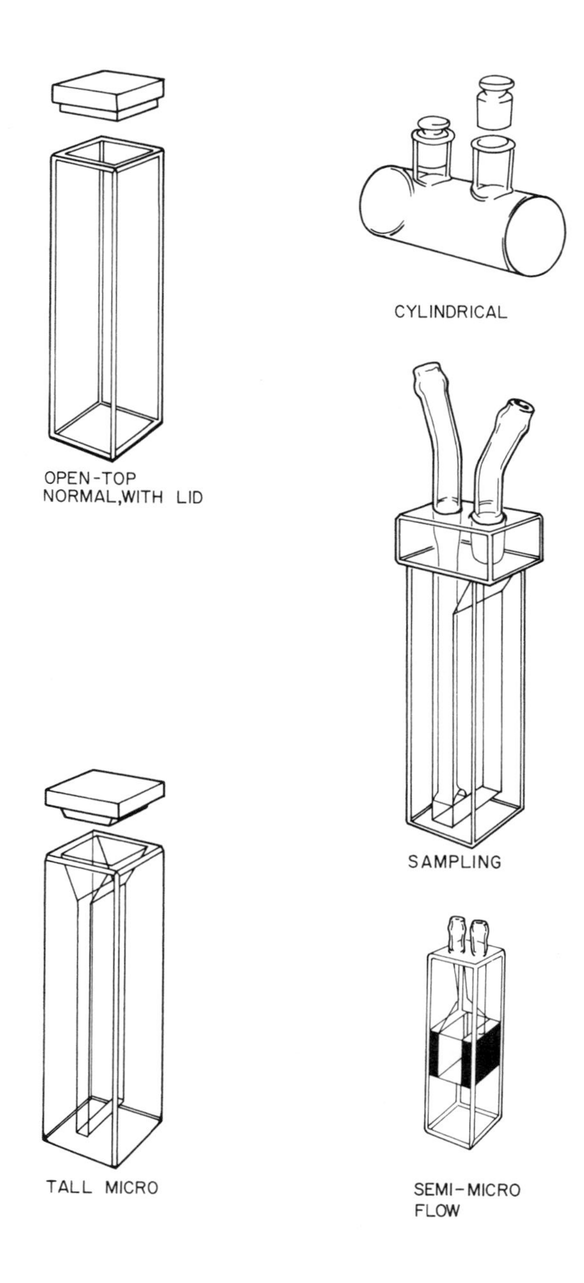

Fig. 2.1b. *A selection of cell types*

These cells can be classified as

Sampling cells

Flow cells

Rectangular cells

Sampling cells are fitted with tubes so that they can be filled and emptied by pressure or vacuum without having to be removed from the instrument. Usually they must be emptied as completely as possible before refilling. This type is often used when a spectrometer has been adapted for a large number of identical determinations involving stable solutions. Many instrument manufactures offer accessories, often called *sippers*, for automatic emptying and filling of sampling cells.

Flow cells are intended for continuous flow operation and are designed so that each sample completely displaces the preceding one. They may be used with a continuously varying sample, as in a chromatography detector unit of the hplc type, or an autoanalyser.

The rectangular cells are simple containers which are filled and emptied manually and may or may not be removed from the instrument for refilling. The popular rectangular 10 mm cells are of this type.

The common types of cells can be obtained in three grades, depending on the optical quality and dimensional tolerances. Grade A cells are of the highest quality and path length tolerances are less than 0.1% eg 10 mm ±0.01 mm. Grade B cells are good quality cells for routine use with tolerances up to 0.5% eg 10 mm ±0.04 mm. Grade C cells are the cheapest and are used as disposable cells for large scale routine work or in elementary teaching labs; dimensional tolerances are often as high as 3% eg 10 mm ±0.3 mm.

Even high quality cells differ a little from each other, and before embarking on accurate quantitative work it is normal to check sets of cells in order to have at least a matching pair, one to be used for the blank or reference solution and the other for the sample(s). The matching should be carried out to ensure that they have the same baseline transmission characteristics, as well as of being of the same materials and possessing identical dimensions.

When you next carry out an analysis in which you use a pair of cells, spend some time making sure you use them correctly. The following comments will help.

When the instrument has been set up with the appropriate wavelength selected, you will set the reading to zero (0.00 absorbance or 100 % transmittance) *before* inserting your cells.

Wash the cells thoroughly, fill both to within 1 cm of the top with the blank solution. Ensure that the outer surfaces are perfectly clean and dry and show no signs of finger marks or smears. Be careful that air bubbles do not start to form on the inner faces of the cells. This often happens if a cool liquid starts to warm up in the cell and the solubility of the dissolved gases decreases.

Now insert your cells in the instrument and check the reading. In our experience pairs of cells are rarely matched and a small, but non-zero reading is observed. If this is less than about 0.02 absorbance then you can set the reading to zero, just as you did prior to inserting the cells. If the reading is > 0.02 absorbance it is very likely that something is wrong and you should check for dirt and air bubbles. Don't be tempted to back-off such a big discrepancy.

SAQ 2.1a

A biochemical enzymatic analysis is being carried out at 340 nm by spectrometric measurements. Indicate which of the following would result in a large (L) and which would result in a small (S) effect on the measured absorbance.

(*i*) The sample becomes cloudy due to poor solubility. L/S

→

SAQ 2.1a (cont.)

(*ii*) The sample is accidently placed in a glass cell instead of a silica cell. L/S

(*iii*) The sample cell is accidently contaminated with propanone. L/S

(*iv*) The tungsten source is used instead of the deuterium source. L/S

(*v*) The pH of the reaction system is not adjusted to the optimum value. L/S

2.2. REAGENTS, COMPLEXATION TECHNIQUES, SOLUTION CONDITIONS

Spectrometric methods of analysis in the uv/visible region show high sensitivity when the material being analysed absorbs strongly somewhere in the wavelength range 200 to 800 nm. Sometimes the component being analysed has its own characteristically strong absorption, but more often it may require the addition of a special reagent to react selectively with the desired component to produce a derivative with the necessary high absorptivity.

Π Do you recall what we mean by high absorptivity?

Let us return to the absorption of copper(II) sulphate in water, referred to in Section 1.1. You may recall that a solution of $CuSO_4$ (0.01 mol dm^{-3}) produced an absorbance of about 0.2 at 800 nm which increased to 0.8 at 600 nm when ammonia was added. The blue colour of the cuprammonium complex shows a reasonably linear Beer's Law plot and has been used for the analysis of samples containing 10 mg or more of copper. However, many other metal ions can also react with ammonia to produce blue coloured complexes, and a more selective colorimetric reagent for copper is diethyldithiocarbamate (DEDC). This reagent produces a characteristic yellow-brown copper complex (with a maximum absorption at 436 nm) which is readily extracted into organic solvents such as chloroform and carbon tetrachloride. It gives absorbance values > 1.0 with microgram quantities of copper ie at concentrations of about 10^{-4} mol dm^{-3}, and its absorptivity is very high.

Π Using the absorbance values and concentrations quoted above calculate approximate values of the molar absorptivity (ε_{max}) for Cu^{2+}, $Cu(NH_3)_4^{2+}$ and the copper-DEDC complex, assuming that the measurements were taken in 10 mm cells.

We rearrange the Beer-Lambert Law to calculate ε from $\varepsilon = A/cl$. The following results are obtained.

	Cu^{2+}	$Cu(NH_3)_4{}^{2+}$	Cu-DEDC
λ_{max} (nm)	800	600	436
ε_{max}($dm^3\ mol^{-1}cm^{-1}$)	20	80	10 000
ε_{max}($m^2\ mol^{-1}$)	2		1000

The reaction of copper ions with DEDC is simple and rapid, and the coloured complex is stable for longer than an hour. By suitable adjustment of the solution conditions the reaction is highly selective for Cu^{2+}. Furthermore, since aqueous solution of the complex can be readily extracted into small volumes of organic solvents, minute traces of copper can be detected.

Thus for any spectrometric analysis it is important to choose the best available reagent and solution conditions for the analyte being determined.

2.2.1 The Ideal Reagent

Reagents for uv/visible analysis can vary from the metal chelating reagents such as 1,10-phenanthroline for iron and DEDC for copper, to enzyme reagents for the analysis of organic and biochemical species. Many enzyme methods are based on recording the change of uv absorption of nicotinamide-adenine dinucleotide (NADH) or the corresponding phosphate (NADPH). Although no reagent performs ideally it is useful to judge the performance of a given reagent against the ideal properties listed below:

(*a*) *Stability in solution.* Some reagents deteriorate in a few hours whilst others ferment or grow moulds on storage. Instability of reagent necessitates freshly prepared solutions and recalibration of the spectrometer for each new batch.

(*b*) *Rapid and reproducible reaction.* For those analytical procedures which require the reaction to go to completion prior to the measurement being made, it is desirable that the reaction be rapid and reproducible. If the reaction is stoichiometric and the product and reagent absorb at different wavelengths, then it is often satisfactory and very convenient to use an excess of the reagent.

The stability of the product is also important and particularly so if measurements are not always made at the same time following initiation of the reaction. The stability may be temperature and/or time dependant.

(*c*) *Reproducible rate of reaction.* Many analytical procedures do not require the reaction to go to completion. Examples are those employing automated methods in which timing can be very accurately controlled, and those such as enzymetric reactions in which the rate of the reaction is measured. In these cases the control of temperature is often very important.

(*d*) *Selectivity or specificity of the reagent-analyte reaction.* This property is important in order that the absorption measured is that for the desired constituent only. This is sometimes difficult to achieve with complexing reagents used for inorganic species. For example, 8-hydroxyquinoline will complex with an extensive range of metal ions under various pH conditions, similarly 2,4-dinitrophenylhydrazine forms derivatives with many aldehydes and ketones. These substances cannot be considered to be specific, but can be used for the determination of a particular *class* of chemical species. In contrast many biochemical reagents, particularly enzymes and immunoassay reagents based on monoclonal preparations, are highly selective.

(*e*) *Solvent compatibility.* The reagent should of course, be soluble in the same solvent used for dissolving the sample and should react to give a complex which is still soluble over a wide concentration range.

(*f*) *Linear calibration.* The product of the reaction should obey the Beer-Lambert Law over a wide range of concentration. This results in linear calibration graphs.

2.2.2 Choice of Reagent

For any given constituent there is probably a number of alternative reagents which could be used in a spectrometric analysis. For example, earlier in this Section we suggested that DEDC was a better reagent for copper than ammonia, mainly because of its higher sensitivity and its better selectivity. However, the blue copper-ammonia complex may be quite suitable for determining the copper content of certain steels, where the concentration levels are appropriate, other reactive elements such as nickel are absent or present only in low concentrations, and the sensitivity requirements are not high. In this situation the copper-ammonia method is rapid, inexpensive and avoids an organic extraction.

There are many reagents available for the analysis of iron ranging from the thiocyanate method for Fe(III) to a variety of chelating agents for Fe(II). The thiocyanate method depends on the formation of a red colour in acid solution and the simplicity of the analytical procedure is one of its main attractions.

Ferrithiocyanate complexes, however, are non-stoichiometric and the colour is unstable. This instability is influenced by the concentration of the reagents, the ionic strength of the solution, and interference due to the presence of ions such as chloride and sulphate. The colour is also a function of pH.

Low concentrations of iron are, therefore, usually determined after initially reducing the iron to the Fe(II) state. This is because this species forms stable complexes with a number of organic reagents (containing suitably placed nitrogen atoms) capable of forming highly coloured and stable complexes.

Some of the best known reagents for the determination of Fe(II) are indicated below:

Reagent	ε_{max}/m^{2}mol^{-1}	λ_{max}/nm
2,2′-dipyridyl	800	522
1,10-phenanthroline	1100	510
4,7-diphenyl-10-phenanthroline	2240	533
2,4,6-tri (2-pyridyl)-1,3,5-triazine	2260	595

The last reagent in this list (TPTZ) was recently selected by the analysts of the British water industry as the best available reagent for the determination of low concentrations of iron. In the report of their investigations they highlighted what they considered were the important features for suitable complexes, as follows:

(*a*) The molar absorptivity (ε_{max}) of the complex should be of the order of 1900 m^{2}mol^{-1} or more to ensure that the required analytical precision can be achieved given the worst precision to be expected of the absorbance measurements.

(*b*) The coloured iron complex must be soluble and stable in aqueous solution so that solvent-extraction procedures are not required.

(*c*) The chromogenic reagent should be commercially available and preferably of similar cost to reagents in common use.

(*d*) The chromogenic reagent must be reasonably selective for iron, and should preferably react with Fe(II) rather than Fe(III).

The TPTZ reagent best fulfilled these conditions and was chosen as the preferred colorimetric reagent for iron in Great Britain. The detailed procedure developed for its use is described in Part 3 of the present Unit. It should be noted that in the USA the prescribed reagent is still 1,10-phenanthroline.

2.2.3 Solution Conditions for Analysis

In the development of an analytical method for the analysis of a species in solution by spectrometry, it is usual to check the accuracy, precision and detection limits of the proposed method with possible variations of solution conditions. Such variations would include:

(*a*) solvent polarity;

(*b*) pH and ionic strength (of aqueous solutions);

(*c*) temperature.

Other factors which might well influence the analysis include:

(*d*) order of addition of reagents;

(*e*) mixing or stirring rate;

(*f*) time allowed for colour development.

The best methods should be independent from most of these variables, but ideal conditions rarely apply and compromise is often necessary.

SAQ 2.2a

Potassium thiocyanate and 1,10-phenanthroline have both been used as reagents for the determination of low concentrations of iron. Both have advantages and disadvantages for this application. Assign as many as possible of the following advantages and disadvantages to the two reagents.

Advantages

(*i*) **Complex formation requires only the addition of the reagent and some acid.**

⟶

SAQ 2.2a (cont.)

(*ii*) The reagent is cheap.

(*iii*) The complex is stable and relatively free from interferences.

(*iv*) The molar absorptivity is over 1000 m^2 mol^{-1}.

(*v*) It is applicable to iron in the Fe(III) state.

Disadvantages

(*vi*) The iron must be reduced to the Fe(II) state.

(*vii*) The complex is non-stoichiometric.

(*viii*) The molar absorptivity is too low to be chosen for water analysis (in the UK).

(*ix*) pH control is important.

(*x*) The complex is sensitive to light, and is relatively unstable.

2.3. CHOICE OF WAVELENGTH AND CALIBRATION DATA

In the previous two Sections we have looked at the factors which have to be considered when a sample is being prepared for spectrometric analysis in the uv/visible region of the spectrum. In this and the following Section we shall consider the optimum instrumental settings necessary to achieve the highest precision possible in the analysis. At this stage we will assume that we already have the sample in solution at the correct concentration level and that the solution is stable and hence the absorbance (or transmittance) is constant with time. We merely have to place the solution in the spectrophotometric cell, choose the appropriate wavelength for measurement, take suitable measurements of the light absorption and deduce the concentration by reference to a set of suitable calibration data. To illustrate the procedure we will go through the analysis of the manganese content of a steel. This procedure is based upon the oxidation of the manganese to the purple coloured permanganate ion in aqueous solution. It is first necessary to prepare a set of calibration solutions and we choose here the set already used in Section 1.2.

2.3.1 Calibration Data Using Permanganate Solutions

The standard solutions of $KMnO_4$ prepared in Section 1.2 have the following concentrations:

Solution	Blank	A	B	C	D	E
Mn Concentration (mg dm^{-3})	0.00	5.00	10.0	15.0	20.0	30.0

Visible absorption spectra of these solutions were recorded over the wavelength range 400 to 650 nm using a double beam spectrometer. These spectra are shown in Fig. 2.3a.

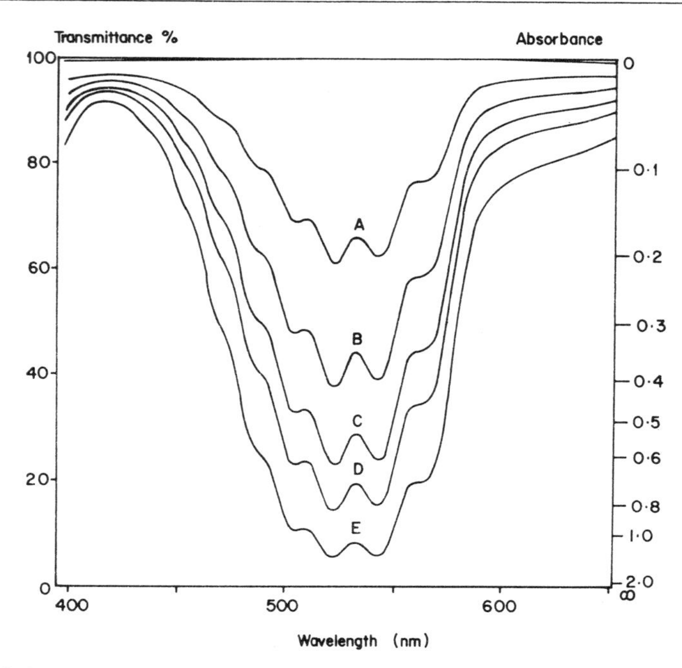

Fig. 2.3a. *Visible absorption spectra of $KMnO_4$ solutions (as in Fig. 1.2c)*

Π As you might expect, this series of solutions shows an increase in the intensity of the purple colour as the Mn content rises. For the strongest solution (E) the width of the absorption band at 50 %T level is about 110 nm; does this cover the wavelength range expected for a purple coloured solution?

Yes. We have covered this point and it is illustrated in Fig. 1.1c.

Prior to recording these spectra, the instrument was set to 520 nm and the zero checked. Two cells, thoroughly washed, were filled with distilled water, and one placed in the sample beam and the other in the reference beam. The instrument balance (zero) was then adjusted to return the pen to zero (100 %T). Only a slight adjustment was necessary, but if the cells had been perfectly matched optically at 520 nm no adjustment would have been necessary. The 'zero' was recorded and represents the spectrum of the water blank used in this experiment. Look carefully and you will see that the blank deviates slightly from zero below about 500 nm, but is within 1 %T of 100 %T over the whole range.

Π Complete the following table of data by making measurements at λ_{max} = 522 nm on the six spectra shown in Fig. 2.3a or use the measurements you made from Fig. 1.2c. Remember $A = \log(100/\%T)$.

		Blank	A	B	C	D	E
λ_{480}	%T	99.9	82.8	69.0	57.5	48.5	38.5
	A	0.000	0.082	0.161	0.240	0.314	0.475
λ_{max}	%T						
	A						

Check the %T value for sample D at 480 nm.

The absorbance data at λ_{max} should be similar to those below.

0.000 0.211 0.421 0.638 0.836 1.236

Π Plot the absorbance for λ_{480} and λ_{max} against concentration of magnanese and draw the best straight line calibration graphs.

Is the Beer-Lambert Law obeyed in both cases (480 and 522 nm)?

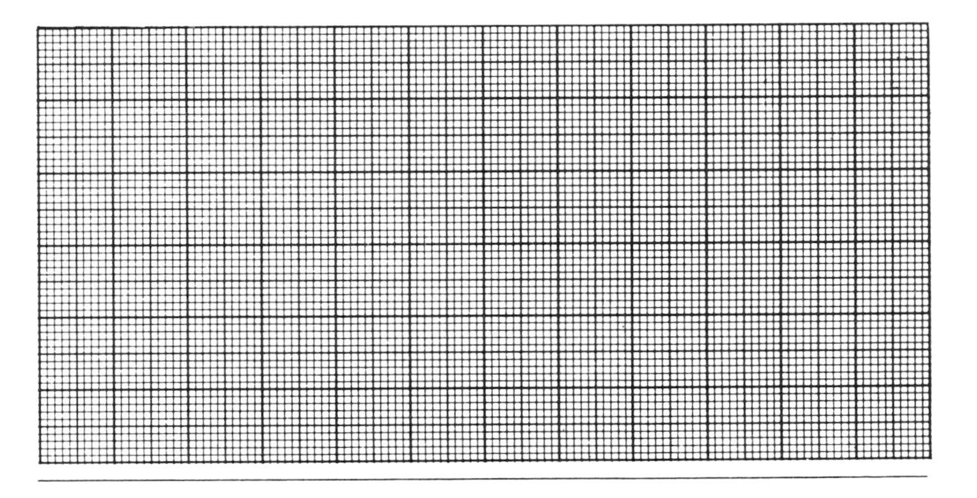

Both calibration graphs are linear over the concentration range 0 to 30 mg dm^{-3}, and pass through the origin, Fig. 2.3b. Hence the Beer–Lambert Law is obeyed at both wavelengths over the full concentration range employed. An additional data point at 25 mg dm^{-3} would have helped in drawing the best line, particular at the high concentration end.

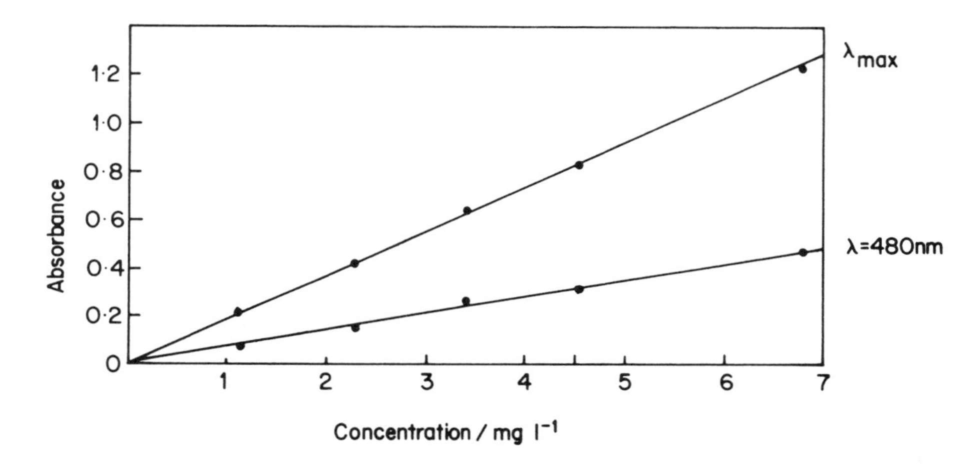

Fig. 2.3b. *Calibration curves for the determination of Mn at 480 nm and 522 nm*

I think you will agree with me that it is more difficult to make measurements at 480 nm then at 522 nm because in the former case we are taking readings from the side of a peak which limits the accuracy. Should a slight error occur in the reading of the wavelength scale, this will result in a considerable error in the corresponding reading taken from the %T or *A* scale. The calibration graph for 480 nm shows slightly greater scatter about the calibration line, and being of lesser slope results in reduced accuracy and sensitivity of the analysis.

Whenever possible measurements should be taken at λ_{max} or at the top of an alternative absorption peak.

SAQ 2.3a

Calculate the absorptivities of $KMnO_4$ using the following data:

A $KMnO_4$ solution at λ_{max} = 522 nm gave an absorbance = 1.236 in a 10 mm cell.

The Mn concentration is 30 mg dm^{-3} [A_r(Mn) = 54.938]

(*i*) Molar Absorptivity, ε_{max}

(*ii*) Absorptivity, $E_{1\%}^{1cm,}$

2.3.2 The Determination of Mn in Steel

1.000 g of steel was dissolved in acid, the manganese oxidised to permanganate, and the volume of the solutions made up to exactly 100 cm^3. The visible absorption spectrum of this solution is shown in Fig. 2.3c.

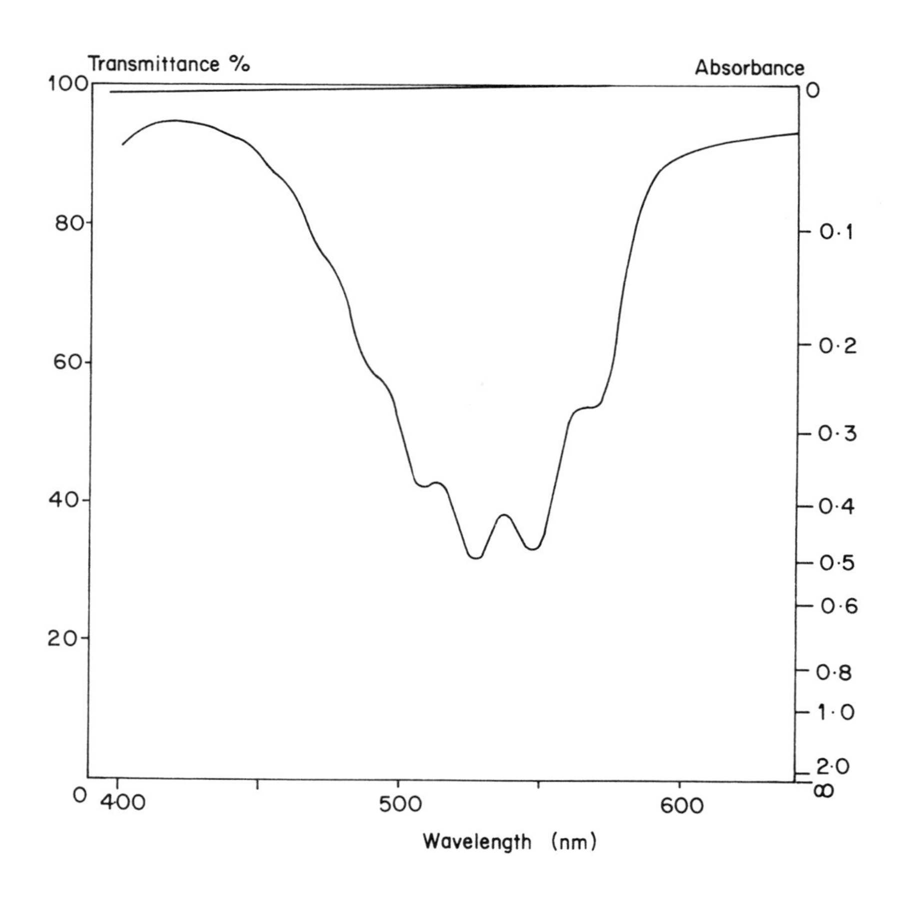

Fig. 2.3c. *Visible absorption spectrum of a solution of $KMnO_4$ of unknown concentration*

Π Read off the transmittance of the steel solution at λ_{max}, and then calculate the corresponding absorbance value.

Yes, the %T is 33 which gives an absorbance of 0.5.

Π Now read off the concentration of Mn in the steel solution using your permanganate calibration curve in Fig. 2.3b.

Your result for the concentration of Mn in the steel solution should be similar to my value of 12 mg dm^{-3}.

The concentration of steel in this solution is 1 g per 100 cm^3, ie 10 g dm^{-3}. Hence the number of grams of Mn in 100 g of steel (ie % w/w) is 120×10^{-3}.

Mn content of steel = 0.12 % w/w.

Throughout this exercise on permanganate solutions we have used a recording uv/visible spectrometer with an ordinate scale linear in transmittance. This has enabled us to consider a number of aspects of analysis such as

(*a*) calculation of absorbance from transmittance values.

(*b*) selection of the analysis wavelength.

(*c*) calibration curves at λ_{max} and other wavelengths.

(*d*) potential errors with respect to quantitive spectrometric measurement.

We could of course have carried out the spectrometric determination of Mn in steel using a number of other types of instrument, such as a simple manual colorimeter operating in the visible region only, with a digital display reading out directly in absorbance. There are many possibilities and I hope that you will carry out such an analysis yourself in the near future if you have not already done so.

2.3.3 The Sample Blank

There is an important aspect of analysis that we have so far ignored, but which can seriously effect accuracy if it is not taken into account. The standards and the sample solutions were of quite different composition and were produced as a result of quite different chemical procedures.

The standard solutions were made up by taking a high purity sample of $KMnO_4$ and dissolving in deionised water.

The steel samples were dissolved in acid, subjected to oxidation, and finally diluted using deionised water.

The most obvious difference in composition is the presence of a range of elements in the solution of the steel, and particularly the presence of a very high concentration of iron.

Π Is it necessary to exactly match the composition of the standards with the sample?

Not always, but we need to find out before publishing our method and results.

One of the simplest ways of testing for accuracy is to purchase an analysed standard sample which has a composition similar to your own sample. The standard can then be analysed and the value(s) checked against the certificate values provided with the standard.

This is fine if such a standard is available, but this is often not the case. When then?

You can employ an alternative method of analysis which is quite different from the one of interest. If the results are in close agreement then you can be quite confident in your accuracy, because it would be very unusual for the errors associated with two quite different methods to be the same.

The evaluation of accuracy is a big subject in its own right and I hope you will take a continuing interest in it. You should read about standard addition methods and inter-laboratory trials whenever you can.

Back to our steel analysis. From where might the errors stem?

(*a*) The reagents used (acid and oxidising reagent) may themselves contain impurities, including Mn itself, which can give rise to absorption at the analysis wavelength of 522 ±5 nm.

(*b*) The constituents of steel, other than Mn, may give rise to coloured products which absorb at the analysis wavelength.

Because this mis-match between the sample and standards is so great in terms of the Fe content, we could test for Fe interference. This can be done by obtaining very high purity iron wire (99.99%) and subjecting it to the dissolution, oxidation and solution procedure. If you do this a yellow solution is obtained which gives zero absorbance at 522 nm. That clears the acid the oxidising reagent and Fe, but leaves possible problems due to other constituent of the sample.

We are not going to pursue this any further, but you should now recognise the need to be on your guard if your results are to be accurate. We have looked at one specific example but the lessons it has taught us are equally applicable to any other sample you may encounter such as river water, blood plasma, sausage meat, and so on.

2.3.4 Wavelength and Absorbance Checks

Most visible spectrometers are supplied with a set of standard filters to be used for checking wavelength calibration over the visible region. An alternative method employed is to use the atomic emission lines from a suitable vapour discharge lamp.

The filters commonly available, are rare-earth (holmium and didymium) oxide glasses which give a series of sharp absorption bands. Some typical wavelength values are shown in the table below;

Wavelength of absorption peaks/nm	
Holmium glass	Didymium glass (neodymium and praesodymium)
241.5 ± 0.2	
279.4 ± 0.3	573.0 ± 3.0
287.5 ± 0.4	586.0 ± 3.0
333.7 ± 0.6	685.0 ± 4.5
360.9 ± 0.8	
418.4 ± 1.1	
453.2 ± 1.4	
536.2 ± 2.3	
637.5 ± 3.8	

However considerable variations occur between different batches of glass, and they are now considered to be unreliable for the calibration of high quality instruments.

The most accurate method of wavelength calibration is by introducing a discharge lamp into the lamp housing of the spectrometer. The most commonly used discharge lamps for this purpose are those of neon and mercury. Some typical wavelengths are:

Neon Emission/nm	Mercury Emission/nm
533.1	
534.1	253.7
540.1	364.9
585.3	404.5
594.5	435.8
614.3	546.1
633.4	576.9
640.2	579.0
667.8	
693.0	
717.4	
724.5	

Apart from these special sources, deuterium emission lines are particularly useful for a quick check, since most uv instruments have a deuterium lamp as a source. The lines used are the red line at 656.1 nm and the blue-green line at 486.0 nm.

An approximate method which may be used in the visible region with an instrument fitted with a tungsten source, is to observe the position where the light passing through the sampling compartment (use a small mirror) appears to be pure yellow in colour, ie neither reddish nor greenish. For most observers with normal colour vision this is observed between 570 to 580 nm.

The *absorbance* scale of an instrument is best checked using one of the recommended solution standards noted below. The use of solution standards combines both operator error and cell errors in one set of measurements and, providing that Beer's Law is obeyed, can be used at any absorbance range within the instrument's capability. The disadvantages of solution standards is that they require careful preparation and show some temperature dependence. Solutions which have been used as standards include;

(*a*) Potassium dichromate in acid solution.

(*b*) Potassium dichromate in alkaline solution.

(*c*) Potassium nitrate.

(*d*) Pyrene in isooctane.

(*e*) Potassium hydrogen phthalate.

(*f*) Nicotinic acid.

Details of the standard values and the wavelength ranges over which the standards can be used are listed in a recent publication on standards in absorption spectrometry, C. Burgess and A. Knowles *Standards in Absorption Spectrometry* (Techniques in visible and ultraviolet spectrometry; Vol.I) Chapman and Hall, London and New York, 1981.

SAQ 2.3b

A solution of potassium permanganate at a concentration of 3.4 mg dm^{-3} Mn transmits 23% at 522 nm and 57.5% at 480 nm. Calculate the effect of a +1% transmittance error on the absorbance at these two wavelengths. Indicate, with reasons, which wavelength is best used for the analysis of permanganate solutions. (522 nm is the λ_{max} for $KMnO_4$, 480 nm is on the side of the absorption band).

2.4 LIMITATIONS IN THE APPLICATIONS OF THE BEER–LAMBERT LAW

The Beer–Lambert Law states that the absorbance of a solution is directly proportional to the path-length and to the concentration of the absorbing species, subject to the following limitations:

(*a*) the solution is of a homogeneous composition,

(*b*) monochromatic radiation is used,

(*c*) the concentration of the absorbing species is low.

In practice we often fail to achieve this direct proportionality and calibration curves are not linear, with the departure from linearity being particularly severe at high concentrations. An example of this is given in Fig. 2.4b which we will be discussing later.

The homogeneous composition of a solution is not normally a problem and we will not consider it further. There are however, four factors which we are going to examine and these include the use of non-monochromatic radiation and concentration effects mentioned above. When we have examined each factor you will be asked what you would do when faced with using a non-linear calibration graph.

2.4.1 Effect of Concentration and pH

As a general rule concentration effects are not usually encountered at concentration < 0.01 mol dm^{-3}. Above this value, however, refractive index changes and the perturbing effects of inter-molecular interactions, or of ionic species, on the charge distribution of the absorbing species can affect the value of the absorption coefficient (absorptivity) and give rise to either positive or negative deviations. In cases where the absorbing compound is involved in a concentration dependant chemical equilibrium such as dimerisation, marked deviations will be observed if the spectral characteristics of the monomer are quite different.

Of particular interest to the analytical scientist is the effect of pH on chemical equilibria, which we make use of so often. For example, we choose a visual indicator for monitoring acid-base titrations whose colour is very dependant on the pH of the solution.

To illustrate, pH dependant equilibria, I have chosen an old favourite – dichromate/chromate. It is of continuing importance in ultraviolet-visible spectrometry because standard solutions of potassium dichromate are used for checking the accuracy of the absorbance scale of instruments.

The equilibrium may be represented as follows:

$$Cr_2O_7^{2-} + H_2O \rightleftharpoons 2CrO_4^- + 2H^+$$

The visible absorption spectrum of a standard solution of $K_2Cr_2O_7$ at different pH is given in Fig. 2.4a

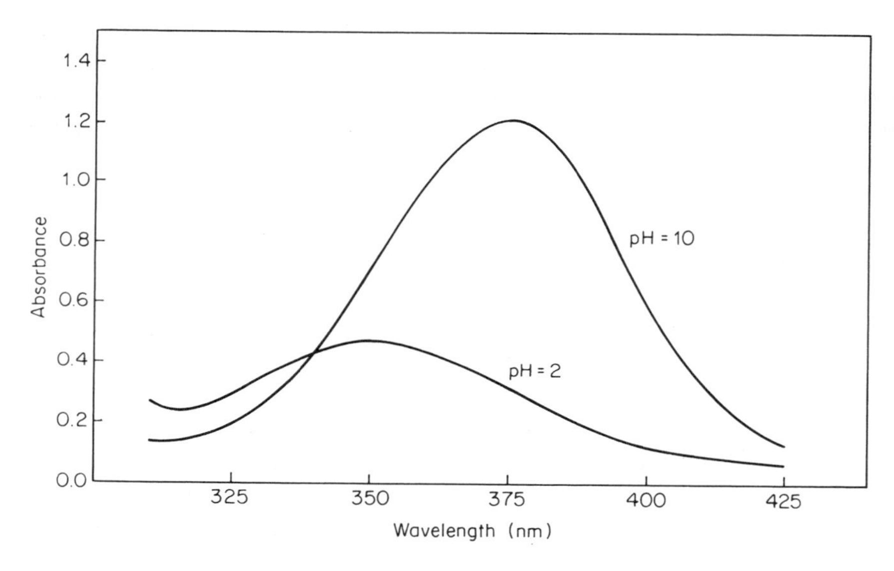

Fig. 2.4a. *Visible absorption spectrum of $K_2Cr_2O_7$ (3×10^{-3} mol dm^{-3}) recorded at pH 2 and 10*

Π Examine Fig. 2.4a and then suggest two possible ways in which you might overcome the problems of the pH dependant equilibrium when preparing a calibration curve for the determination of chromium(VI) using standard solutions of potassium dichromate.

One of the ways is chemical and the other spectral.

Yes, buffer all solutions. If the pH is buffered at say less than pH = 3 you would make your measurements at 348 nm. An alternative would be to make all measurements on unbuffered solutions at the isosbestic point, because the absorbance is independent of pH at this wavelength. 340 nm in this case.

Note that when we have two absorbing species which are interconvertible, as are dichromate and chromate in our example, then their spectra may overlap. The wavelength at which overlap occurs is called the *isosbestic point*, and the absorbance at this wavelength is independant of the position of equilibrium, and depends only on the total amount of the substances present.

2.4.2 Instrument and Spectral Bandwidths

In a spectrometer the ideal monochromator would enable the selection of radiation of any single wavelength within a given range, but in practice this situation is never achieved and instead of a single wavelength the 'monochromated' beam actually has a spread of values on either side of the required value. This instrument bandwidth corresponds to the width of the transmitted band at half the maximum transmittance value.

Modern uv/visible instruments usually have adjustable bandwidth settings, typically in the range 0.5 to 8 nm. The higher bandwidths are used where samples are strongly scattering or absorbing, ie where the signal is low. But wide bandwidths do have an effect of absorption curve profile that needs to be appreciated (see below). Conversely, if the fine structure in a spectrum is to be resolved then narrow bandwidths are required, subject to the signal/noise requirements.

As indicated above the magnitude of the instrument bandwidth selected for measurement can influence noise levels, spectral resolution and absorbance values. For quantitative analysis it is the influence on absorbance values which is of most importance. The effect is a function of the ratio of the instrument bandwidth (IBW) to the spectral bandwidth (SBW) of the sample being measured.

Typical factors for the effect on absorbance values are:

Ratio IBW/SBW	0.1	0.25	0.5	1.0	1.5
Absorbance factor	1.0	0.96	0.87	0.66	0.55

This table shows that for a given instrument bandwidth, the effect is most noticeable for materials with a sharp absorption peak in the uv/visible region (eg benzene and its derivatives).

The data also shows that the instrument bandwidth needs to be 10 x less than the spectral bandwidth of the species being measured if accurate absorbance values are to be obtained. This is important because if the ratio IBW/SBW is not less than 0.1 the Beer–Lambert Law will not be obeyed and negative deviations will occur. Measurements made on the sloping side of an absorption band are affected even more seriously by relatively large instrument bandwidths.

2.4.3 Stray Radiation

When we set a monochromator to pass radiation of a particular wavelength we wish to prevent radiation at a range of higher and lower wavelengths from reaching the detector. If the instrument has a stray light level of 0.1% we can only stop 99.9% of the unwanted radiation reaching the detector. Let us find out how good this and other levels are and the effect that stray light has on the absorbance values we obtain.

We have already considered an equation relating absorbance to % transmittance.

$$A = \log \frac{100}{\%\text{T}} \qquad (1.5)$$

Let us modify it to enable us to calculate the effects of stray light.

If S is the % stray light our equation becomes

$$A = \log \frac{(100 + S)}{(\%T + S)} \tag{2.1}$$

This equation tells us that absorbance values are reduced by the presence of stray light and that the reduction becomes more serious at high values of absorbance. This is illustrated in Fig. 2.4b, which shows the apparent negative deviation from the Beer-Lambert Law as a result of stray radiation.

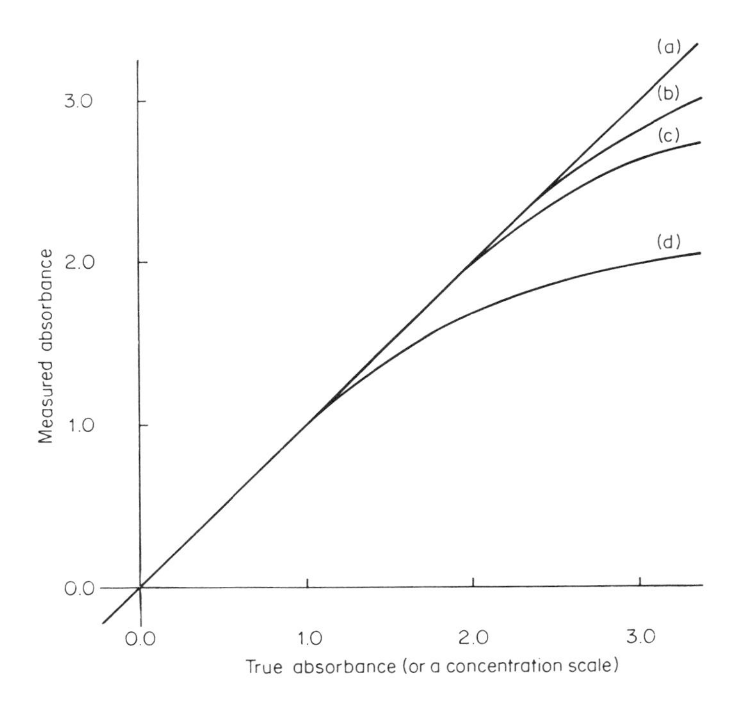

Fig. 2.4b. *Deviations from the Beer-Lambert Law due to stray radiation (S)*

(a) $S = 0\%$ *(b)* $S = 0.05\%$

(c) $S = 0.1\%$ *(d)* $S = 1.0\%$

In addition the presence of stray radiation can also give rise to distortions in the shape and position of absorption bands (Fig. 2.4c).

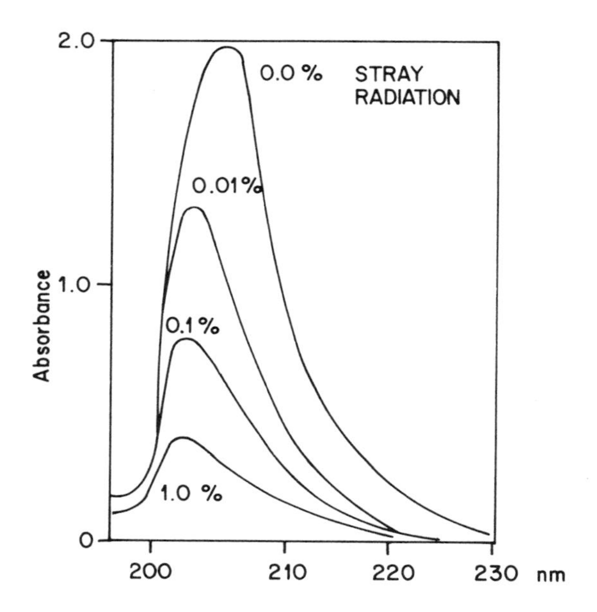

Fig. 2.4c. *Stray radiation effect on the absorption spectrum of maleic acid in ethanol*

The main source of stray radiation in most spectrometers is usually the dispersing element in the monochromator, although scattering from other optical surfaces and deposited dust increases markedly with the age of the instrument. The effect of stray radiation is most noticeable in the low wavelength region of spectrometers particularly below 220 nm, where source energy and detector sensitivity are lowest and also where the transmission characteristics through the optics are decreasing. The stray radiation characteristics are usually quoted as a percentage of the transmittance signal at say 220 nm, and for a new instrument should be less than 0.1%.

2.4.4 Non-linear Calibration Graphs

What are the problems we face when the Beer-Lambert plot is non-linear? Can we use non-linear calibration graphs and obtain accurate and precise analytical results?

Let us begin to answer these questions by drawing our own curve. Fig. 2.4d is an incomplete calibration curve in which six data points have been plotted. Complete the figure by drawing the best curved line to fit the points.

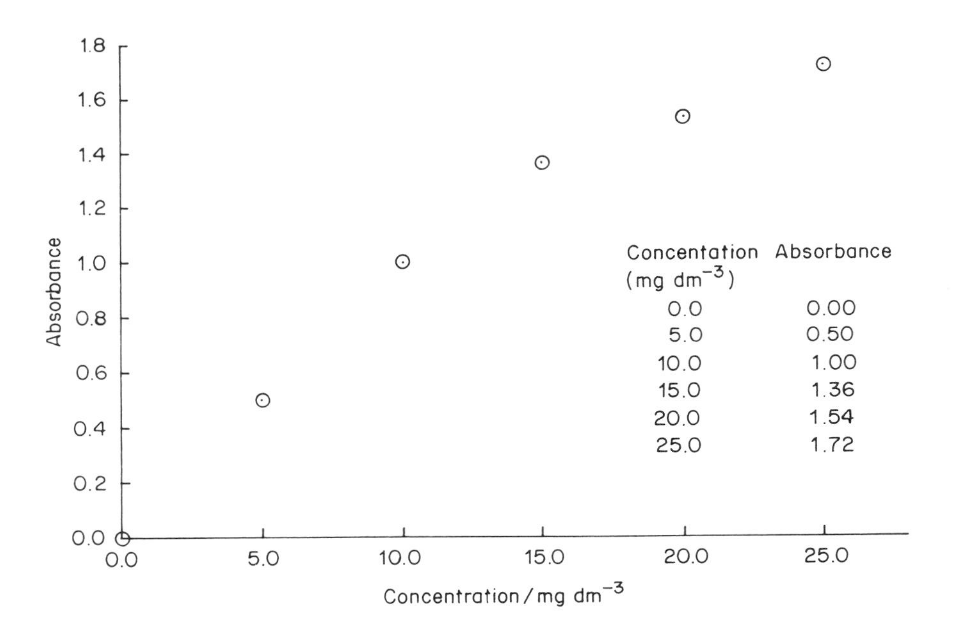

Fig. 2.4d. *A partially constructed calibration curve*

If you managed to draw a perfectly smooth narrow line by having available a sharp pencil and a flexi-curve ruler, you would have been left with the difficulty of deciding exactly *where* to draw the line.

Π Would more data points in the region of 10 to 25 mg dm^{-3} have been helpful?

Yes. It is not easy to accurately define the curve in this region. We could also have done better by calculating the best curve using a third order polynomial and you would have done this if you had access to a micro computer with appropriate software. Even so a calculated curve would also leave something to be desired because of the limited number of data points that is available in this case.

Π A solution of unknown concentration gave an absorbance of 1.45. Use your calibration curve to determine the concentration of this solution. Do you think your answer is likely to be inaccurate?

Probably, because it depends very strongly on the exact positioning of your curve and on how accurately you interpolated between the absorbance and concentration axes. Our value is 17.1 mg dm^{-3}

Π It is obviously bad practice to use the region of a calibration graph where the curvature is pronounced. If you were faced with this problem of a solution giving an absorbance corresponding to a particularly non-linear region of the curve, what should you do?

Accurately dilute the solution so that the results fall in a linear region of the graph. Calibration curves for analyte species with small absorption coefficients can often deviate from the Beer-Lambert Law even at fairly low concentrations, and dilution of the sample is no longer advantageous. Are we justified in using the curve? When the non-linearity is not severe the graph itself can be drawn and read with accuracy. Are we justified, therefore, in using this curve when the basic Law of qualitative spectrometry is not strictly obeyed? That depends on whether or not the deviation from the Law, which is evident from measurements on the standard solutions, is exactly the same for the solution of the sample being analysed.

We have already discussed problems that can arise when the composition of the sample and calibrating standard solutions are not the same (matrix matched). You should now appreciate that it is possible to have two solutions of identical analyte concentration giving rise to different absorbance values because the composition due to non-analyte species is not the same.

We started this section by asking two questions. The first is rephrased in the following SAQ.

SAQ 2.4a

List two of the problems that occur when calibration data do not obey the Beer-Lambert Law over the concentration range of interest.

The second question was – can we use non-linear calibration graphs and obtain accurate and precise analytical results?

You should be able to do so if you are careful. You are now aware of the problems and how they might be reduced to acceptable levels.

SAQ 2.4b

Absorbance measurements are to be made on a series of aqueous solutions of copper sulphate (0 to 0.1 mol dm^{-3}, λ_{max} = 800 nm, ε_{max} = 20 m^2 mol^{-1}). The instrument being used covers the region 350 to 900 nm and has a stray light specification of 1%.

Unfortunately the detector has low sensitivity at wavelengths above 700 nm. Indicate what each of the following options has to offer in our search for high precision and simplicity of analysis.

(*i*) Increase instrument bandpass from 2 to 8 nm to improve signal strength,

(*ii*) Decrease the sample path length from 10 to 5 mm in order to diminish the effects of stray radiation,

(*iii*) Modify the analysis procedure and form the copper-ammonia complex with λ_{max} = 600 nm and ε_{max} = 80 m^2 mol^{-1},

(*iv*) Modify the analysis procedure and form the Cu-DEDC complex with λ_{max} = 430 nm and $\varepsilon_{max} \approx 10^4$ m^2 mol^{-1}.

SAQ 2.4b

Summary

Quantitative determinations in uv/visible spectrometry are dependant upon reliable methods of sampling materials, the treatment required to get all the component into solution and any chemical reaction used to obtain a coloured or highly uv active species. The uv/visible absorption for the sample material has to be compared with a calibration graph previously prepared from a series of standard solutions of known concentration in order to establish the concentration of the analyte in the sample.

Objectives

You should now:

- understand the importance of the various steps involved in sampling and in preparing that sample for analysis;
- have a knowledge relating to solvents and the effects of acidity which may determine the quality of the absorption spectra obtained from samples in solution;
- appreciate the important features in the choice of a reagent in producing a stable derivative with a strong absorption peak;
- be able to select the most suitable wavelength from a spectrum for use for quantitative measurements;
- have learnt the importance of wavelength and absorbance checks on spectrometers.

3. Spectrometric Determinations

In Parts 1 and 2 we considered the hardware used for spectrometric analysis in the uv/visible, ie the solvents and cells, the reagents and solution additives, the instruments and their operational characteristics and the factors which limit the precision of methods of spectrophotometric analysis. In this part we consider the procedures used in analysis, the manner in which we can manipulate the data, and the range of applications available to us.

3.1 TYPICAL ANALYTICAL PROCEDURES

Let us return to the question posed at the beginning of Part 2, ie how would you go about determining the iron content of water using a colorimetric method and a visible spectrometer?

Π Can you now list the five or more steps which need to be specified? You can – good! If not, then refer back to that section and remind yourself.

But how much detail can or should you remember?

Obviously an analytical technician engaged in repetitive and routine analysis will quickly learn the precise details of amounts to weigh, volumes of reagent solutions to add and the sampling and operational conditions for the instrument being used. However, it is not the intention of this course to supply these details, instead we hope you will learn to appreciate the items of detail you need to know and, most importantly, where you can find the necessary information.

In your working environment you will be aware of the sources of information and analytical procedures available immediately to you – but what about information available elsewhere? There are, of course, standard methods laid down by the British Standards Institute (BSI) and the American Society for Testing of Materials (ASTM standards), as well as methods specified within a particular industry (eg the food or water industries). Whatever your sources, they need to provide you with all the information you require in order to carry out the analysis with the optimum accuracy and precision. However, not everything can be learnt from books or the analytical literature - you need to have some 'hands-on' experience of the particular analysis being attempted before you can expect to achieve reliable results in which other people will place confidence.

In this Part we will start by looking at contrasting methods for the analysis of single components, such as iron and glucose. After this we shall consider the details of an investigation into the development of a method for ultra trace level analysis of iron. Next we will look at the principles involved in the analysis of mixtures, and which leads to the question of the manipulation of spectrophotometric data. Finally we will summarise the application range of spectrophotometric methods covering the ultraviolet and visible region.

3.1.1 Spectrophotometic Methods for Iron in Water

Standard methods for the analysis of drinking waters are defined by government bodies in most countries. In the U.K. one of the joint technical committees of the Department of Environment and the National Water Council is responsible for the provision and publication of recommended methods for the analysis of water. However, because both needs and the availability of equipment vary widely, a selection of methods may be recommended for a single component. In the USA the analytical procedures are given in the publication *Standard Methods for the Examination of Water and Waste Water* which is published jointly by the American Public Health Association, the American Water Works Association and the Water Pollution Control Federation.

In the latest edition of this publication two methods are listed for the determination of iron in water. Of the two methods, one uses a procedure known as atomic absorption.

This is said to be relatively easy to carry out with both the precision and accuracy being superior to those of the second method which uses a colorimeter. However, the colorimetric method involving 1,10-phenanthroline requires much less expensive instrumentation and is simple and reliable and is still used extensively. In the US publication the determination of iron content by 1,10-phenanthroline is detailed under six headings.

1. General Discussion (principles, interference, concentration levels).

2. Apparatus.

3. Reagents.

4. Procedure.

5. Calculation.

6. Precision and Accuracy.

The corresponding standard method in the U.K. *Iron in Raw and Potable Waters by Spectrophotometry (1977)* published by Her Majesty's Stationery Office (HMSO) specifies the use of 2,4,6-tripyridyl-1,3,5-triazine (TPTZ) as the preferred colorimetric reagent for iron. This publication details the method of analysis under 14 headings.

1. Performance Characteristics of the Method.
2. Principle.
3. Interferences.
4. Hazards.
5. Reagents.
6. Apparatus.
7. Sample Collection and Preservation.
8. Sample Pretreatment.
9. Analytical Procedure.
10. Measurement of Absorption.
11. Preparation of Calibration Curve.
12. Change in Concentration Range of the Method.
13. Checking the Accuracy of Analytical Results.
14. References.

As this Unit is concerned with the practical aspects of analysis itself, it is assumed that sample collection and pretreatment (7 and 8) have been correctly carried out. We are particularly interested in the analytical procedure, and Sections 9 and 10 are reproduced below. Read through them carefully and note that guidance is given on:

(*a*) temperature control (9.1),

(*b*) time factor (9.2),

(*c*) cell size and measurement wavelength (9.3),

(*d*) the importance of blank determinations (9.4 and 9.7),

(*e*) compensation for sample colour and turbidity (9.6),

(*f*) calculation of the results (9.9-9.11).

Iron in Raw and Potable Waters by Spectrophotometry (1977 version)

Note: Throughout this method iron is expressed as the element (Fe)

1 Performance Characteristics of the Method

(For further information the determination and definition of performance characteristics see another publication in this series).

1.1	Substance determined	All forms of iron (see Sections 2 and 8).			
1.2	Type of sample	Raw and potable waters.			
1.3	Basis of method	Reduction of iron to the ferrous state and subsequent reaction with 2, 4, 6-tripyridyl-1, 3, 5-triazine to form a coloured complex whose concentration is measured spectrophotometrically.			
1.4	Range of application (a)	Up to 1 mg/l.			
1.5	Calibration curve (a)	Linear to 2 mg/l at 595 nm.			
1.6	Total standard deviation	Iron concentration (mg/l)	Total standard deviation (mg/l)		Degrees of freedom
	1.6.1 Without pretreatment:	0.150	0.005	(a) (d)	15
		0.250	0.004	(a) (c)	17
		0.400	0.003–0.008	(b) (c)	6–9
		1.000	0.011	(a) (c)	19
	1.6.2 With pretreatment:	0.250	0.013	(a) (c)	13
		0.334	0.019	(a) (d)	8
		1.000	0.013	(a) (c)	13
1.7	Limit of detection				
	1.7.1 Without pretreatment (b)	0.003–0.015 mg/l (with 5 to 10 degrees of freedom)			
	1.7.2 With pretreatment (a)	0.06 mg/l (with 7 degrees of freedom).			
1.8	Sensitivity (a)	1.0 mg/l gives an absorbance of approximately 1.25			
1.9	Bias (a)	No important sources of bias were detected.			
1.10	Interferences (a)	None of the substances tested caused appreciable errors except commercial polyphosphate (See Section 3).			
1.11	Time required for analysis (a)	The total analytical and operator times are the same. Typical times for 1 and 10 samples are approximately 45 and 60 minutes excluding any pretreatment time.			

(a) These data were obtained by the Water Research Centre (Medmenham Laboratory)[1] using this method and a spectrophotometer with 40-mm cells at 595 nm.
(b) These data were obtained from an interlaboratory calibration exercise in which 5 laboratories took part.[2]
(c) These data were obtained using distilled water spiked with the stated iron concentration.
(d) River Thames water.

9 Analytical Procedure

READ SECTION 4 ON HAZARDS BEFORE STARTING THIS PROCEDURE

Step	Experimental Procedure	Notes
	Analysis of samples	
9.1	Add 40.0 ± 0.5 ml of the well mixed sample to a 50-ml calibrated flask. Adjust the temperature of the sample, if necessary, to between 15 and 30 °C (notes e and f).	(e) If the sample contains polyphosphate, see Section 3 note (g). (f) See Section 12 for concentration range.
9.2	Add to the flask, swirling after each addition, 2.0 ± 0.1 ml of 10% m/V hydroxylammonium chloride solution, 2.0 ± 0.1 ml of 0.075% m/V TPTZ solution, and 5.0 ± 0.2 ml of acetate buffer solution. Dilute with water to the mark, stopper the flask, and mix the contents well (notes g and h). Allow to stand between 5 minutes and 2 hours.	(g) If a batch of samples is to be analysed, each reagent can be added to all samples before adding the next reagent. (h) If pretreatment has been used see Section 8.1.
9.3	Meanwhile set up the spectrophotometer (see Section 6.2) according to the manufacturer's instructions. Adjust the zero of the instrument with water in the reference cell. Measure the absorbance (see Section 10) of the well mixed solution at 595 nm using 40-mm cells (note i). Recheck the instrument zero. Let the absorbance of the sample be S.	(i) Other sizes of cells may be used but the performance characteristics quoted in Section 1 would no longer apply.
	Blank determination (if pretreatment not required)	
9.4	A blank must be included with each batch (eg up to 10 samples) of determinations for which pretreatment was not required using the same batch of reagents as for samples. Add 0.80 ± 0.05 ml of 5M hydrochloric acid and 39 ± 1 ml of water to a 50-ml calibrated flask and adjust the temperature to between 15 and 30 °C.	
9.5	Carry out steps 9.2 and 9.3, let the absorbance of the blank be B.	

Step	Experimental Procedure	Notes
	Compensation for colour and turbidity in the sample (note d)	
9.6	A sample compensation solution must be included with each sample for which a colour/turbidity correction is necessary using the same batch of reagents as for samples. Carry out steps 9.1 to 9.3 but omitting addition of the TPTZ reagent. Let the absorbance of the sample compensation solution be S_1.	
	Determination of iron in the water used for the blank (notes j and k)	
9.7	Add 1.60 ± 0.05 ml of 5M hydrochloric acid and 29 ± 1 ml of water to a 50-ml calibrated flask and adjust the temperature to between 15 and 30 °C. Add to the flask, swirling after each addition, 4.0 ± 0.1 ml of 10% m/V hydroxyammonium chloride solution, 4.0 ± 0.1 ml of 0.075% m/V TPTZ solution and 10.0 ± 0.2 ml of acetate buffer solution. Dilute with water to the mark, stopper the flask, mix the contents well and carry out step 9.3. Let the absorbance be D.	(j) This determination is not needed if the iron content of the water used for the blank is known or is negligible (Section 13.3). (k) All reagents must be from the same batch as for the samples.
9.8	The absorbance due to iron in 50 ml of water W is given by: $W = 2B - D - C$ where C = absorbance of sample cell when it and the reference cell are filled with water. Calculate the iron concentration in the water C_w from 0.8 W (note l) and the calibration curve. (See Section 11).	(l) The factor 0.8 allows for the fact that the calibration curve is for 40 ml samples whereas W was obtained for an effective 50 ml sample.
	Calculation of results	
9.9	Calculate the apparent absorbance due to iron in the sample, R, from $R = S - B$ or, when a correction for colour/turbidity is made $R = S - B - S_1 + C$	
9.10	Determine the apparent iron concentration, C_a, in the sample from R and the calibration curve. (See Section 11).	
9.11	Calculate the iron concentration in the original sample, C_r, from $C_r = 1.02\,(C_a + C_w)$ mg/l (note m)	(m) The factor 1.02 allows for the dilution of the sample by the acid into which it was collected. (See Section 7).

With all methods of analysis you should be aware of the scope of the method as well as sources of error and the precision and accuracy. These are specified in Section 1, of the Paper. They show that the reliability of the procedure has been evaluated in several laboratories in order to clearly establish its limitations.

SAQ 3.1a

In many colorimetric analyses it is sufficient to correct the sample absorbance (S) for the blank reading (B), and to read off the component concentration from the Beer's Law calibration graph. This is implied in the above procedure for iron in water when the apparent absorbance is given by:

$$R = S - B$$

Indicate whether using the above simple procedure the factors below would result in a high (H), low (L) or correct (C) value for the iron content of the water being analysed. If you have insufficient information to make a judgement use the code (I).

(*i*) The temperature of the sample dropped to 15 °C before measurement.

(*ii*) The instrument was found to have a 1 nm calibration error.

(*iii*) The sulphuric acid reagent was found to contain 1 mg dm^{-3} iron.

(*iv*) The deionised water used for the blank was found to contain 0.1 mg dm^{-3} iron.

(*v*) Only 1/10th of the quantity of hydroxylammonium chloride reagent was added (0.2 cm^3 instead of 2.0 cm^3).

(*vi*) The final solution for measurement looked slightly cloudy.

SAQ 3.1a

3.1.2 The Uv/visible Determinations of Glucose

Spectrophotometric methods for the determination of sugars, and particularly glucose, in the blood of diabetic patients have been much studied by clinical and biochemical analysts, and the determination of sugars is also of interest in the food industry. Although titrimetric methods were originally used, colorimetric methods were developed as early as 1920 and spectrometry applied to the measurement of the developed colour in the 1940's.

These early methods relied on the ability of glucose to reduce copper(II) or ferricyanide solutions, with reagents being added to the system to produce characteristically coloured complexes with the reduced copper(I) and ferricyanide products. Even now a modern automated method of determining glucose in blood samples is based upon measurement of the loss of intensity of the yellow colour of the ferricyanide reagent using a photoelectric colorimeter fitted with a flow-through cell.

However, these reduction methods are susceptible to errors arising from the presence of other reducing species (eg glutathione in blood). Enzyme reagents are much more specific and can be designed to be based on the formation of a coloured product or, as mentioned in Part 2, on the change in the uv spectrum of nicotinamide-adenine dinucleotide (NAD) as it is quantitatively reduced to NADH. The change in the spectrum is illustrated in Fig. 3.1a.

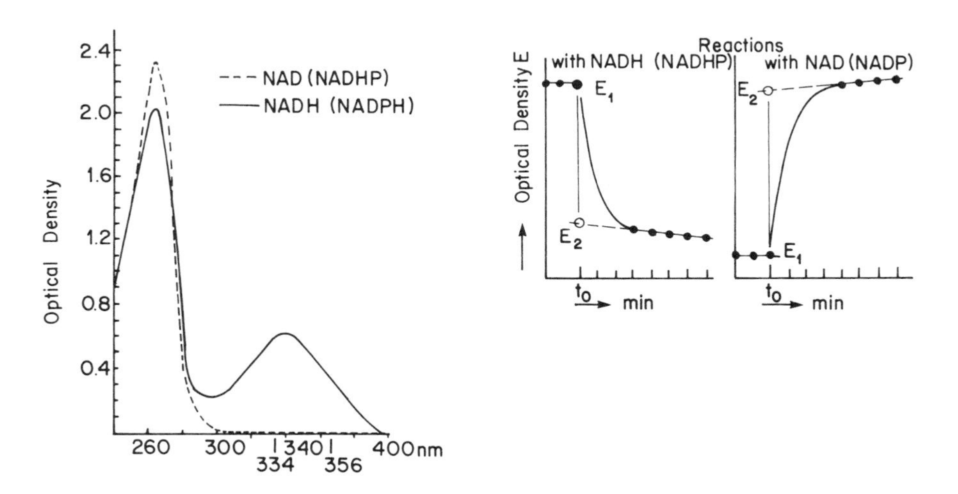

Fig. 3.1a. *Changes in uv absorption at 340 nm of NAD and NADH when used in enzymatic analyses*

∏ An example of an enzymatic analytical procedure called 'uv-method for the determination of sucrose and glucose in food-stuffs' is given below. At the moment simply make a written note of the section headings.

Sucrose/Glucose

food analysis
boehringer mannheim

UV-method

for the determination of sucrose and glucose in foodstuffs
Cat. No. 139041

Test-Combination
for 40 sucrose or glucose determinations

Principle

The glucose concentration is determined before and after enzymatic hydrolysis.

Determination of Glucose before Inversion:

At pH 7.6, the enzyme hexokinase (HK) catalyses the phosphorylation of glucose by adenosine triphosphate (ATP) (1). In the presence of glucose-6-phosphate dehydrogenase (G6P-DH) the glucose-6-phosphate (G6P) produced is specifically oxidized by nicotinamide-adenine-dinucleotide-phosphate (NADP) to gluconate-6-phosphate with the formation of reduced nicotinamide-adenine-dinucleotide phosphate (2).

$$(1)\ \text{Glucose} + \text{ATP} \xrightarrow{\text{HK}} \text{G6P} + \text{ADP}^*$$

$$(2)\ \text{G6P} + \text{NADP}^+ \xrightarrow{\text{G6P-DH}} \text{gluconate-6-phosphate} + \text{NADPH} + \text{H}^+$$

The NADPH formed in this reaction is stoichiometric with the amount of glucose and is measured by the increase in absorbance at 334, 340 or 365 nm.

Enzymatic Inversion

At pH 4.6, sucrose is hydrolysed by the enzyme β-fructosidase (invertase) to glucose and fructose (3).

$$(3)\ \text{Sucrose} + H_2O \xrightarrow{\beta\text{-fructosidase}} \text{glucose} + \text{fructose}$$

The determination of glucose after inversion (total glucose) is carried out simultaneously according to the principle outlined above.

The sucrose content is calculated from the difference of the glucose concentrations before and after enzymatic inversion.

Each Test-Combination Contains

1. Bottle 1 with ca. 7.2 g of lyophilisate, consisting of: triethanolamine buffer – pH 7.6, NADP – 102 mg, ATP – 257 mg, magnesium sulphate, stabilizers.
2. Bottle 2 with 1.1 ml of enzyme suspension, consisting of: hexokinase – ca. 320 U, glucose-6-phosphate dehydrogenase – ca. 160 U.
3. Bottle 3 with ca. 0.5 g of lyophilisate, consisting of: citrate buffer – pH 4.6, β-fructosidase – ca. 720 U, stabilizers.

Preparation of Solutions

1. Dissolve contents of bottle 1 in 43 ml redist. water.
2. Use contents of bottle 2 undiluted.
3. Dissolve contents of bottle 3 in 10 ml redist. water.

Stability of Solutions

Solution 1 is stable for 4 weeks when stored at +4 °C, or for 2 months at −20 °C, resp. The contents of bottle 2 are stable for 1 year when stored at +4 °C. Solution 3 is stable for at least 4 weeks when stored at +4 °C, or for 2 months at −20 °C, resp.

Procedure

Wavelength[1]: 340 nm, Hg 365 or Hg 334 nm
Glass cuvette[2]: 1 cm light path
Temperature: 20–25° C
Final volume: 3.02 ml

Read against air (without a cuvette in the light path) or against water.
Sample solution: 5–150 µg of sucrose and glucose/cuvette[3]

Pipette into cuvettes	blank	glucose sample	sucrose sample
solution 3 sample solution	0.2 ml –	– 0.1 ml	0.2 ml 0.1 ml
mix, allow to stand for 15 min at 20–25° C. Mix in			
solution 1 redist. water	1.0 ml 1.8 ml	1.0 ml 1.9 ml	1.0 ml 1.7 ml
read absorbances of the solutions after approx. 3 min (A_1). Start reaction by addition of			
suspension 2	0.02 ml	0.02 ml	0.02 ml
mix, wait for the end of the reaction (approx. 10–15 min), and read absorbances of the solutions (A_2). If the reaction has not stopped after 15 min, continue to read the absorbances at 5-min intervals until the absorbance increases constantly over 5 min.			

If the absorbance at A_2 increases constantly, extrapolate the absorbances A_2 to the time of the addition of suspension 2.

Calculate the absorbance differences A_2–A_1 for both blank and sample. Subtract the absorbance difference of the blank from the absorbance difference of the sample. The difference $\Delta A_{total\ glucose}$ (from the sucrose sample) and $\Delta A_{glucose}$ (from the glucose sample) yields $\Delta A_{sucrose}$.

$$\Delta A_{sucrose} = \Delta A_{total\ glucose} - \Delta A_{glucose}$$

Calculations

According to the general formula for calculating the concentrations, the equation is:

$$c = \frac{V \times MW}{\varepsilon \times d \times v \times 1000} \times \Delta A\ [\text{g/l}]$$

where V = final volume [ml]
v = sample volume [ml]
MW = molecular weight of the substance to be assayed
d = light path [cm]
ε = absorption coefficient of NADPH at
340 nm = 6.3 [l × mmol⁻¹ × cm⁻¹]
Hg 365 nm = 3.5 [l × mmol⁻¹ × cm⁻¹]
Hg 334 nm = 6.18 [l × mmol⁻¹ × cm⁻¹]

It follows for

sucrose

$$c = \frac{3.02 \times 342.30}{\varepsilon \times 1 \times 0.1 \times 1000} \times \Delta A_{sucr} = 10.34 \times \frac{\Delta A_{sucr}}{\varepsilon}$$

[g sucrose/l sample solution]

glucose

$$c = \frac{3.02 \times 180.16}{\varepsilon \times 1 \times 0.1 \times 1000} \times \Delta A_{gluc} = 5.441 \times \frac{\Delta A_{gluc}}{\varepsilon}$$

[g glucose/l sample solution]

If the sample has been diluted on preparation, the result must be multiplied by the dilution factor F.

Further instructions

1. **Performance of Assay**

The amount of sucrose and glucose in the cuvette should range between 5 µg and 150 µg (measurement at 365 nm) or 5 µg and 80 µg (measurement at 340, 334 nm), respectively. The sample solution must therefore be diluted sufficiently to yield a sugar concentration between 0.05 and 1.5 g/l or 0.05 and 0.8 g/l respectively.

* ADP = Adenosine-5'-diphosphate
1 The absorption maximum of NADPH is at 340 nm. With spectrophotometers, measurements are made at the absorption maximum; when using spectrum line photometers equipped with a mercury vapour lamp, measurements are carried out at a wavelength of 365 nm or 334 nm.
2 If desired disposable cuvettes may be used instead of glass cuvettes.
3 See instructions for performance of the assay and sample preparation.

Dilution Table

estimated amount of sucrose + glucose per litre measurements at 340 or 334 nm	365 nm	dilution with water	dilution factor F
< 0.8 g	< 1.5 g	-	1
0.8–8.0 g	1.5–15.0 g	1 + 9	10
8.0–80 g	15.0–150 g	1 + 99	100
> 80 g	> 150 g	1 + 999	1000

If the amount of sucrose + glucose is below 0.05 g/l, the sample volume to be pipetted into the cuvette can be increased (up to 2.0 ml for the glucose sample, and up to 1.8 ml for the sucrose sample). The volume of water to be added must then be reduced so as to obtain the same total volume for the sample and blank in the cuvettes. The different sample volume v must be taken into account in the calculations.

If the amount of sucrose estimated is below 0.2 g/l, the incubation period specified in the assay scheme (in order to allow β-fructosidase to split sucrose) may be reduced from 15 min to 5 min.

If the glucose: sucrose ratio is greater than 10 : 1, the excess glucose must first be destroyed by means of GOD/Catalase so that the concentration of sucrose may be accurately determined. Proceed as described under "honey", section 2.3.b.

2. **Preparation of Sample**

2.1. **Liquid Foodstuffs**

Determination of Sucrose and Glucose in Fruit Juices

Filter turbid juices and dilute sufficiently to obtain a sucrose and glucose concentration of approx. 0.1-1.5 g/l. The diluted sample solution can also be used for the assay if it is colored. When *strongly* colored juices are being used *undiluted* for the assay, owing to their low sucrose content, they must first be decolorized. In this case the following procedure is used:

Mix 10 ml of juice and approx. 0.1 g of polyamide powder of polyvinylpolypyrrolidone, stir for 1 min and filter rapidly. The clear, slightly colored solution is used for the assay.

Determination of Sucrose and Glucose in Wine

The wine should be pretreated as described for "fruit juices". Even strongly colored sweet wines need not be decolorized.

Determination of Sucrose and Glucose in Beer

To remove the carbonic acid, stir approx. 5–10 ml of beer in a beaker for about 30 seconds with a glass rod. The largely CO_2-free sample can be used undiluted for the assay.

Determination of Sucrose in Sweetened Condensed Milk

Weigh ca. 1 g of sample accurately into a 100 ml volumetric flask, add about 60 ml redist. water and incubate for 15 min at ca. 70 °C; shake from time to time. For protein precipitation, add 5 ml of diluted Carrez-I-solution (3.60 g potassium hexacyanoferrate-II, $K_4[Fe(CN)_6] \cdot 3H_2O$/100 ml), 5 ml of Carrez-II-solution (7.20 g of zinc sulphate, $ZnSO_4 \cdot 7H_2O$/100 ml) and ca. 10 ml of NaOH (0.1 mol/l), adjust to room temperature and fill up with water to the mark, filter. Use the clear, possibly slightly opalescent solution for the assay. Dilution according to Dilution Table.

2.2. **Solid Foodstuffs**

Mince foodstuffs in an electric mixer, meat grinder or mortar. Weigh in the well mixed sample and extract with water (heated, if necessary, to 60° C). Quantitatively transfer to a volumetric flask and dilute to the mark with water. Filter, and use the clear solution, which may be diluted if necessary, for the assay.

Determination of Sucrose in Chocolate

Weigh in accurately approx. 1 g of chocolate, finely grated, into a 100 ml volumetric flask, add approx. 70 ml of water, and heat in a water bath at 60–65° C for 20 min. Shake from time to time. After the chocolate has been completely suspended, cool and dilute to the mark with water. To separate the fat, keep in a refrigerator for at least 20 min. Filter the cold solution through a filter paper which has been moistened with the solution. Discard the first few ml of the filtrate. Transfer 10 ml of the clear filtrate to a 50 ml volumetric flask and dilute to the mark with water (dilution factor F = 5). Use the diluted solution for the assay.

2.3. **Pasty Products**

Determination of Sucrose and Glucose in Jam

Homogenize approx. 10 g of jam in an electric mixer. Accurately weigh in ca. 0.5 g of the homogenized jam into a 100 ml volumetric flask, mix with water, and dilute to the mark. Filter through a rapidly filtering fluted paper. Discard the first 5 ml of the filtrate. Remove 1 ml of the clear filtrate obtained and dilute with 9 ml of water (dilution factor F = 10).

Use the diluted solution for the assay.

Determination of Sucrose and Glucose in Honey

Thoroughly stir the honey with a spatula. Remove approx. 10 g of the viscous (or crystalline) honey, heat in a beaker for 15 min at about 60° C, and stir occasionally with a spatula (there is no need to heat liquid honey). Allow to cool. Accurately weigh in approx. 1 g of the liquid sample into a 100 ml volumetric flask. Dissolve at first with only a small portion of water, and then dilute to the mark.

Determination of Glucose

Dilute the 1 % honey solution 1 + 9 and use for the assay.

Determination of Sucrose

a) If the estimated sucrose content in the honey lies between 5 and 10 %, dilute the 1 % solution 1 + 2 and use for the assay.

b) If the estimated sucrose content in the honey lies between 0.5 and 5 %, the sucrose determination must be preceded by substantial destruction of the excess glucose. This is achieved by means of glucose oxidase (GOD):

$$\text{Glucose} + H_2O + O_2 \xrightarrow{\text{GOD}} \text{gluconate} + H_2O_2$$

The hydrogen peroxide is destroyed by catalase:

$$2\,H_2O_2 \xrightarrow{\text{catalase}} 2\,H_2O + O_2$$

Reagents

Glucose oxidase (GOD), Cat. No. 105139[4];
Catalase, Cat. No. 106810[4];
Triethanolamine hydrochloride, Cat. No. 127426[4];
$MgSO_4 \cdot 7H_2O$; NaOH, 4 mol/l.

Preparation of Solutions for 10 Determinations

Enzyme solution:

Dissolve 5 mg GOD in 0.75 ml redist. water, add 0.25 ml catalase, and mix.

Buffer solution:

Dissolve 5.6 g triethanolamine hydrochloride and 0.1 g $MgSO_4 \cdot 7H_2O$ in 80 ml water, adjust to pH 7.6 with sodium hydroxide (4 mol/l), and fill up with water to 100 ml.

Stability of Solutions

The enzyme solution must be prepared freshly.

The buffer solution is stable for at least 4 weeks when stored at 4° C.

Procedure for Glucose Oxidation

Pipette into 10-ml volumetric flask:	
buffer solution	2.0 ml
sample solution (up to ca. 0.5 % glucose)	5.0 ml
enzyme solution	0.1 ml
Pass a current of air (O_2) through the mixture for 1 hr; check the pH with indicator paper during the oxidation and, if necessary, neutralize the acid formed with NaOH.	

To inactivate the enzymes GOD and catalase, heat the volumetric flask in a boiling water bath for 15 min, allow to cool, and dilute to the mark with water. Mix and filter, if necessary. Use 0.5 ml of the clear solution for the determination of sucrose. Determine the residual glucose in a parallel assay and subtract as usual.

BOEHRINGER MANNHEIM GMBH
Biochemica

880 832.15.3

4 Available from Boehringer Mannheim GmbH

The method given is one detailed in a brochure of a company supplying reagents for the enzymatic analysis of food products. Let us now check the headings you wrote down and examine the detail under each one, as we go along.

(*a*) The *analytical principles* are outlined at the beginning of the instructions, and the *types of samples* which can be analysed are detailed in Section 2 of the *further instructions*.

(*b*) The *amounts of sample* are given in Section 1 of the further instructions, with the *sample preparation* methods being detailed in Section 2.

(*c*) The *reagents* (11 of them!) are fully specified along with the methods of *preparing solutions* (I-VI). You should note that the keeping qualities (shelf lives) of these solutions are also specified - an important point, particularly with biochemical reagents.

(*d*) The *analytical procedure* includes details of;

(*i*) volumes of reagent solutions and the final volume,

(*ii*) the composition of the blank,

(*iii*) the temperature of mixing and time before measurement,

(*iv*) the cell size and type, and wavelength of measurement.

Π Can you indicate how this procedure differs in principle from that detailed for the iron determination in Section 3.1.1?

First of all the mixing is done directly in the cell or cuvette and the total volume is not an integer number of ml (3.14 ml!). Secondly the procedure involves the measurement of differences in absorbance, (or optical density E - note the difference in terminology and symbols from those used throughout this Unit). Thirdly, time is an important factor; the reaction products do not form immediately and a method of coping with a non-stable final reading is specified. Did you spot all three points?

(*e*) *Calculation of the analytical result* utilises previously determined values of the molar absorptivity: ε, rather than the preparation of a set of calibration standards. This, of course, limits the *accuracy of the calculated result* since ε values are subject to instrumental variations. But also note that the procedure allows measurements to be taken at wavelengths away from λ_{max}. It is surprising that there is no indication of the standard deviation expected from the method adopted, and certainly no mention of the larger variation expected when disposable cells are used!

Finally the details given show that the method was adapted from a *standard text* on *Methods of Enzymatic Analysis*

Π Can you suggest what advantages the brochure method will have over that specified in a textbook?

Brochure methods of this type usually specify exact details which are selected to give optimum result with the reagents being supplied. Textbooks often only give general experimental details which have to be adapted for any particular analysis being attempted.

SAQ 3.1b

The success of the enzymatic determination of glucose by the method described in Section 3.1.2 is implicit in the statement that 'NADPH is stoichiometric with the amount of glucose and is determined by means of its absorption at 334, 340 or 365 nm'. Why do you think three wavelengths are specified and what are the implications in terms of the instrumental precision achievable?

SAQ 3.1b

3.1.3 Investigating a Spectrometric Procedure

The detailed specifications of the solution conditions in a given analytical procedure are usually those which result in rapid and quantitative formation of the absorbing species, in minimising deviations from Beer's Law (Section 2.4), and in the suppression of interferences from other components present. However, it is only by reading an account of the development of the method or by long experience that you will come to appreciate which aspects of the procedure are the critical ones in respect of achieving good reproducibility (high precision) and accuracy.

The following details of the determination of ultratraces of iron in analytical grade reagents has been abstracted from the *Analyst* (December, 1976 Vol. 101,pp.974-981). This is a British journal devoted to accounts of investigations into analytical procedures, and the article referred to has been chosen for the following reasons.

(*a*) It illustrates how the high sensitivity of a specially developed colorimetric procedure is achieved by the use of a complex which is concentrated by solvent extraction prior to measurement of the absorption (absorbance or percent transmittance).

(*b*) The investigation includes the comparison of reducing agents, the testing of optimum pH conditions, and even the influence of time of shaking the reaction mixture.

(*c*) It reminds you that there are always questions to be asked about purity of reagents, and the stability of the complex formed.

This paper is typical of a detailed investigation of a proposed analytical procedure. It is written in the standard style used in the chemical literature with a *summary* of the work and the main achievements, and then the *introduction* which deals with the background information based on previous work in the area, and so on. In the reproduction which follows we have chosen to omit the *introduction* in order to concentrate on the analytical procedure and its assessment. For the same reason we have omitted the *list of references* and the *conclusions.* The *conclusions* are of course of great interest, but you will not miss the main ones if you read the *summary* at the beginning of the paper.

As you now read through the paper try and appreciate the nature and reasons for each of the investigations, and note their consequences in the standard procedure adopted. For example, why is the sample and reagent mixture shaken for 30 minutes and why is hydroxylammonium chloride the chosen reducing agent?

974 *Analyst, December, 1976, Vol. 101, pp. 974–981*

Spectrophotometric Determination of Trace Amounts of Iron in Pure Reagent Chemicals by Solvent Extraction as the Ternary Complex of Iron(II), 4-Chloro-2-nitrosophenol and Rhodamine B

Kyoji Tôei, Shoji Motomizu and Takashi Korenaga

Department of Chemistry, Faculty of Science, Okayama University, 3-1-1, Tsushimanaka, Okayama-shi, 700, Japan

Trace amounts of iron in pure reagent chemicals have been determined by solvent extraction - spectrophotometry with 4-chloro-2-nitrosophenol and Rhodamine B. The ternary complex of iron(II), 4-chloro-2-nitrosophenol and Rhodamine B was extracted quantitatively into toluene at about pH 4.8. The absorbance of the organic phase was measured in a glass cell of 10-mm path length at 558 nm. The apparent molar absorptivity of the ternary complex in the organic phase was 9.0×10^4 l mol^{-1} cm^{-1} at 558 nm. The ternary complex was very stable and not decomposed by addition of EDTA. By using the above procedure, trace amounts of iron (10^{-7}–10^{-4}%) in alkali metal salts, alkaline earth metal salts, ammonium salts, acids and bases, etc., were determined. The standard deviations of the determinations were 1–3%.

Procedure

A 20-ml volume of the sample solution (or a smaller portion if a large amount of iron was present) was transferred by pipette into a stoppered test-tube and 1 ml of buffer solution (pH 4.8) containing 10% of hydroxylammonium chloride added, followed by 1 ml of 5×10^{-3} M Rhodamine B solution and 1 ml of 1×10^{-3} M 4-chloro-2-nitrosophenol solution in that order. The solution was well mixed and 5 ml of toluene were added. The tube containing the mixture was shaken horizontally in a shaker at a rate of 280 oscillations min^{-1} for 30 min and allowed to stand for 10 min. The absorbance of the toluene phase was measured in a glass cell of 10-mm path length at 558 nm against the reagent blank.

Results and Discussion

Effect of pH

Iron was extracted quantitatively into toluene at pH 3.9–5.3 as the ternary complex iron(II) - 4-chloro-2-nitrosophenol - Rhodamine B. In this work, toluene was selected as the extraction solvent and the pH was adjusted to 4.8 so as to ensure greater accuracy.

Shaking Time

As shown in Fig. 1, when the reagent was added as a solution in water the ternary complex was quantitatively extracted by shaking for 20 min. If a toluene solution containing 4-chloro-2-nitrosophenol was added, the extraction time was at least 50 min. The use of an aqueous solution of 4-chloro-2-nitrosophenol was chosen with a shaking time of 30 min.

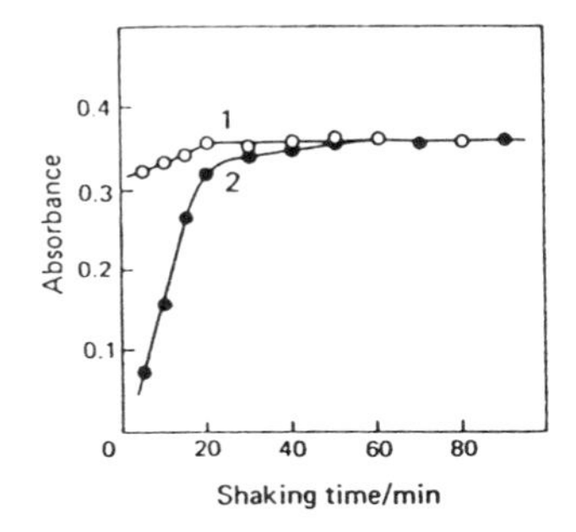

Fig. 1. Absorbance *versus* shaking time: 1, added as an aqueous solution of 4-chloro-2-nitrosophenol; and 2, added as a toluene solution of 4-chloro-2-nitrosophenol. Sodium chloride (manufacturer C, grade c, see Table II), 4.00 g per 20 ml.

Reducing Agent

Any iron in pure reagent chemicals was almost entirely present as the trivalent metal ion. The reduction of iron(III) to iron(II) by nine reducing agents was studied. They were examined at concentrations of 0.5 and 5% and the results are given in Table I. At a concentration of 0.5%, all of these reducing agents except ascorbic acid could reduce iron(III) quantitatively to iron(II). Hydroxylammonium chloride and sulphate and sodium thiosulphate were effective at the 5% level, but 4-chloro-2-nitrosophenol was reduced and decomposed by ascorbic acid, sodium sulphite and sodium dithionite. From these results, it was decided that hydroxylammonium salts were suitable for reducing iron(III) to iron(II) and 0.5% of hydroxylammonium chloride was used in this work.

TABLE I

SELECTION OF REDUCING AGENT

An amount of ammonium iron(III) sulphate containing 2.23 μg of iron was added to each solution. All absorbance values were measured against a reagent blank.

	Absorbance	
Reducing agent	0.5% of reducing agent	5% of reducing agent
$NH_2OH.HCl$	0.796	0.799
$(NH_2OH)_2.H_2SO_4$	0.791	0.801
$(NH_2)_2.H_2SO_4$	0.804	0.108
Ascorbic acid	0.368	—*
Hydroquinone	0.791	0.275
Na_2SO_3	0.777	—*
$Na_2S_2O_3$	0.799	0.810
$Na_2S_2O_4$	0.783	—*
NaH_2PO_2	0.780	0.305

* 4-Chloro-2-nitrosophenol was decomposed.

Absorption Spectra

In Fig. 2, the absorption spectra of 4-chloro-2-nitrosophenol, Rhodamine B, the ternary complex iron(II) - 4-chloro-2-nitrosophenol - Rhodamine B and its reagent blank in toluene are shown. The wavelengths of maximum absorption of the ternary complex occur at 420, 558 and 700 nm. The absorption maxima at 420 and 700 nm are the absorbances of 4-chloro-2-nitrosophenol itself and the green complex of iron(II) and nitrosophenol, respectively. The absorbance of the reagent blank at 558 nm is about 0.087. The absorption spectra of sodium and potassium chloride solutions, etc., are very similar to those of distilled water. The absorbance was therefore measured at 558 nm.

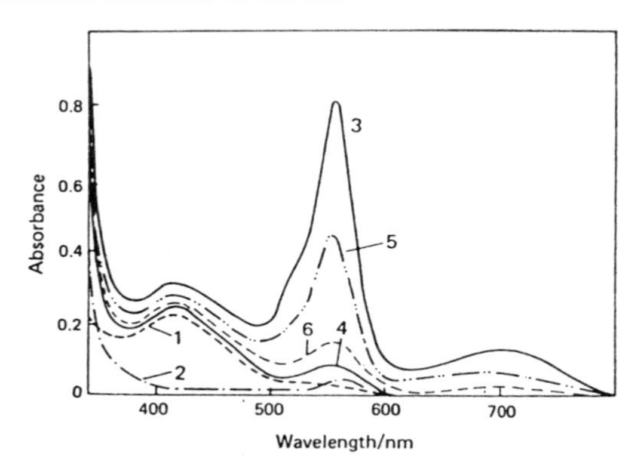

Fig. 2. Absorption spectra in toluene: 1, 2 × 10^{-4} M 4-chloro-2-nitrosophenol; 2, 1 × 10^{-3} M Rhodamine B; 3, ternary complex iron(II) - 4-chloro-2-nitrosophenol - Rhodamine B [8 × 10^{-6} M iron(II); 2 × 10^{-4} M 4-chloro-2-nitrosophenol; 1 × 10^{-3} M Rhodamine B]; 4, reagent blank, no iron added; 5, sodium chloride (manufacturer D, grade a, see Table II), 4.03 g per 20 ml; and 6, potassium chloride (manufacturer A, grade a, see Table II), 2.02 g per 20 ml. pH 4.8; reference, toluene.

Interference of Other Ions

It has been shown that the amounts of cobalt, nickel, copper and tin(II) ions generally present in pure reagent chemicals such as alkali metal salts, alkaline earth metal salts, acids and bases do not interfere with this method.

Monovalent anions having large molecular volumes, such as iodide, perchlorate and thiocyanide ions, could be extracted into the toluene phase as ion pairs with Rhodamine B and the absorbances of the reagent blank were greater than that of distilled water. Masking agents for iron, such as fluoride ion, interfere with the determination.

Effect of Concentration of Sample Solutions

Volumes of 100 ml of each sample solution were placed in 200-ml separating funnels and buffer solution containing hydroxylammonium chloride, Rhodamine B solution, 4-chloro-2-nitrosophenol solution and toluene were added. Iron was removed from these solutions as the ternary complex and the aqueous phase, which was free from iron, was retained and diluted as required.

A known amount of iron(II) was added to the diluted aqueous phase and the effect of the concentration of salts in the sample solutions was examined; the results are shown in Fig. 3. The results for iron in the presence of sodium and ammonium salts were not affected by the concentration of these salts. Potassium salts caused considerable interference when present at levels greater than 10% and aluminium salts at levels greater than 2%. This procedure enabled the most favourable conditions for the determination of iron in various salts to be selected.

Reagent Blank and Calibration Graph

The reagent blanks were examined by using sample solutions from which the iron had been removed. The absorption spectra of these reagent blanks were almost the same as that of distilled water to which 0.2 ml of hydrochloric acid (1 + 1) per 100 ml had been added.

By adding known amounts of iron(II) ions to these iron-free sample solutions, calibration graphs were prepared and found to be straight lines with slopes almost the same as that of the calibration graph obtained by using distilled water. Accordingly, the latter graph was used to determine iron in all sample solutions.

Stability of Iron in Sample Solutions

As shown in Fig. 4, iron in 20% sodium chloride solution was not stable and the absorbance was found to decrease gradually and become zero after 5 h. When 20% sodium chloride solution to which 0.2 ml of hydrochloric acid (1 + 1) per 100 ml had been added as soon as the sample solution had been prepared was used, the absorbance increased gradually, became constant after 5 h and remained at this level for at least 1 week. When the 20% sodium chloride solution to which hydrochloric acid (1 + 1) had been added was heated at 80–90 °C for about 10 min constant absorbance values were obtained and they were identical with those obtained by using the solution that had been allowed to stand for 5 h.

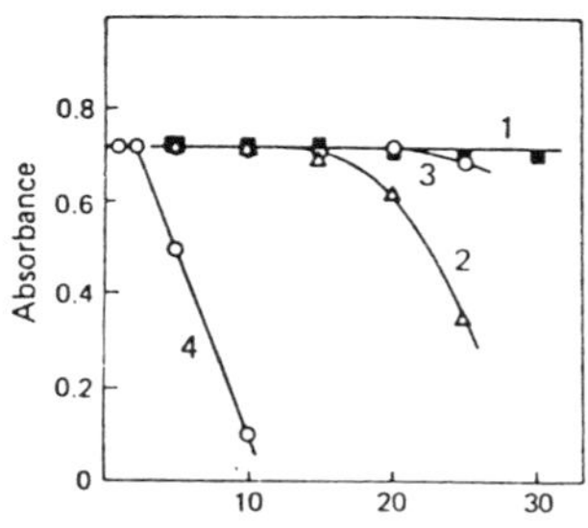

Fig. 3. Absorbance *versus* concentration of sample solutions: 1, sodium chloride; 2, potassium chloride; 3, ammonium chloride; and 4, aluminium nitrate. 8×10^{-6} M iron(II); pH 4.8; reference, reagent blank.

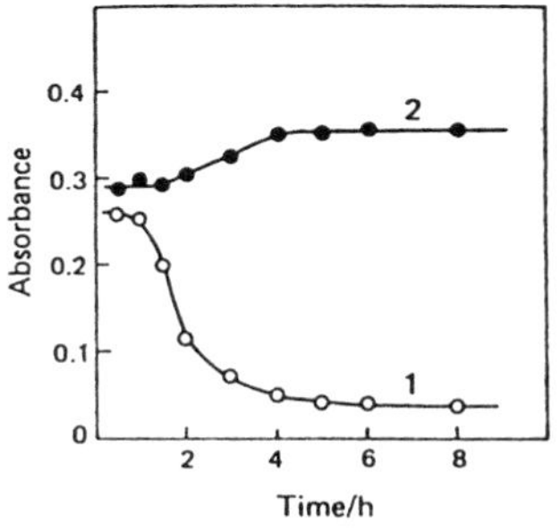

Fig. 4. Stability of iron in sodium chloride solution: 1, no hydrochloric acid (1 + 1) added; and 2, 0.2 ml of hydrochloric acid (1 + 1) per 100 ml added. Sodium chloride (manufacturer C, grade e, see Table II), 4.00 g per 20 ml.

SAQ 3.1c

> The molar absorptivity (ε_{max}) for the ternary complex of iron(II), 4-chloro-2-nitrosophenol and Rhodamine B is quoted as 9.0 × 10^3 m^2mol^{-1} at 558 nm. This is only a factor of four (4x) times greater than that obtained with the iron(II), tripyridyltriazine (TPTZ) complex, yet the typical iron concentration levels measured are at least ten times lower (50 μg dm^{-3} against 500 μg dm^{-3}). How is this achieved?

3.2 BINARY AND MULTICOMPONENT SYSTEMS

The vast majority of the many thousands of uv/visible methods which have been developed are designed for the determination of a single component in a sample. However, one of the advantages of spectrometry over visual colorimetry is the ability to handle the analysis of mixtures of absorbing species. In recent years the analysis of complex mixtures has usually involved chromatographic methods, particularly GC and HPLC, but these methods are often time consuming and expensive in reagents and solvents. With the development of computer-linked spectrometry it is possible to undertake direct uv/visible analysis of mixtures using curve-fitting techniques. Such applications will increase in popularity as the hardware and software for the computer-assisted analysis are developed.

3.2.1 Principles of the Analysis of Binary Systems

In a binary absorbing system the absorption spectra of the two components can overlap to different extents as illustrated in Fig. 3.2a.

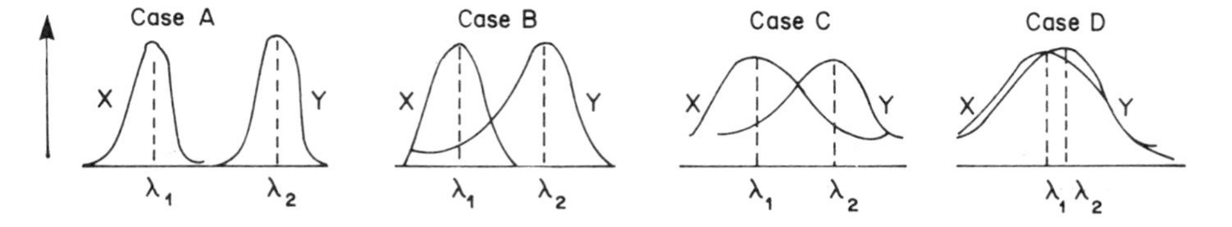

Fig. 3.2a. *Absorption Profiles for Binary Mixture Analyses*

Case A is the situation where the curves show no overlap and the two components (X and Y) are determined directly at wavelengths λ_1 and λ_2. This situation is common in the infrared region, but less common in the uv/visible due to the very broad absorption bands which normally occur.

Case B arises from partial overlap of the absorption spectra. Whilst measurements at λ_2 give Y directly, any attempt to measure the absorbance of X includes some contribution due to the tail from the absorption of Y.

Case C is the general case of overlapping absorption curves but with absorption maxima sufficiently displaced to allow a fairly accurate analysis by the method discussed below.

Case D is for materials with closely matching absorption curves and is not amenable to the analysis discussed below.

For the *Case C* situation involving a mixture (M) of X at concentration c_X and Y at concentration c_Y we assume that the absorbances at any wavelength are additive, such that for the mixture

$$A_M = A_X + A_Y \tag{3.1}$$

$$A_M = a_X c_X l + a_Y c_Y l \tag{3.2}$$

In order to use measurements of the absorbance (S_M) of the mixture to calculate the concentrations c_X and c_Y in the mixture, we need to take measurements at two different wavelengths (say λ' and λ'') then

$$A'_M = a'_X c_X l + a'_Y c_Y l \tag{3.3}$$

$$A''_M = a''_X c_X l + a''_Y c_Y l \tag{3.4}$$

These two equations can be solved for c_X and c_Y provided we have the values of the four constants (absorptivities), a'_X, a''_X, a'_Y, a''_Y.

Π Can you state in words what further assumptions are implied by Eqs. 3.1 and 3.2 and hence the significance of the *a* and *l* values?

Eq. 3.1 not only implies that the absorbances of the components in the mixture are additive, but also that they continue to obey the Beer–Lambert Law after being mixed together. From your knowledge of the Beer–Lambert Law you will know that *a* is the absorptivity for the substance(s) at the selected wavelength and *l* the cell path length used for the mixture.

The solution of Eqs. 3.3 and 3.4 is best illustrated by an example:

Example of the Analysis of a Binary Mixture

A mixture of *o*-xylene and *p*-xylene is to be analysed by uv spectrometry in the range 240 nm to 280 nm. The absorption spectra of their solutions in cyclohexane are illustrated in Fig. 3.2b.

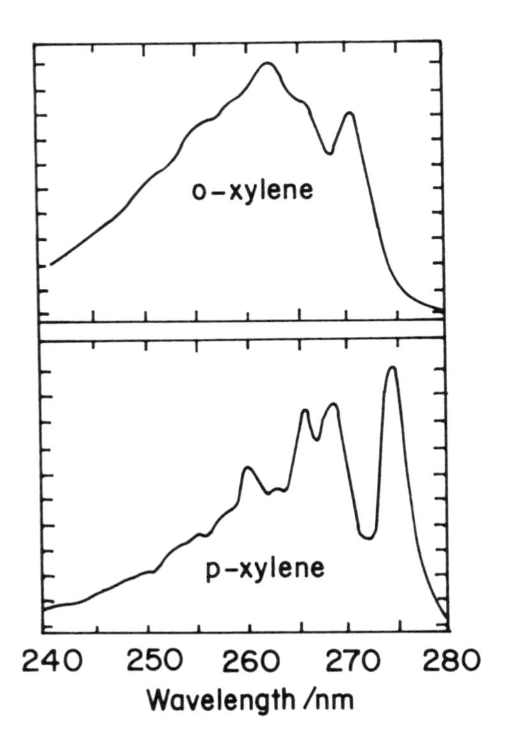

Fig. 3.2b. *Ultraviolet absorption spectra of xylenes in cyclohexane solution*

The spectra show that *o*-xylene has absorption maxima at 263 nm and 271 nm and *p*-xylene at 275 nm. The optimum wavelengths for analysis of the mixture are 271 nm and 275 nm; the 271 nm peak of *o*-xylene overlaps with a low absorption of the *p*-xylene at this wavelength, and the 275 nm peak of *p*-xylene overlaps with a low absorption due to the *o*-xylene.

Measurements on the single components and the mixture at these two wavelengths are:

	271 nm (λ')	275 nm (λ'')
o-xylene (0.40 g dm^{-3})	0.90	0.10
p-xylene (0.17 g dm^{-3})	0.34	1.02
mixture	0.47	0.54

The figures in the table are absorbances measured in 10 mm silica cells.

The values of the absorptivities are obtained from the single component solutions as follows:

For *o*-xylene (component X)

$$271 \text{ nm}(\lambda')\ a'_X = \frac{0.90}{0.40} = 2.25 \text{ dm}^3 \text{ g}^{-1} \text{ cm}^{-1}$$

$$275 \text{ nm}(\lambda'')\ a''_X = \frac{0.10}{0.40} = 0.25 \text{ dm}^3 \text{ g}^{-1} \text{ cm}^{-1}$$

For *p*-xylene (component Y)

$$271 \text{ nm}(\lambda')\ a'_Y = \frac{0.34}{0.17} = 2.0 \text{ dm}^3 \text{ g}^{-1} \text{ cm}^{-1}$$

$$275 \text{ nm}(\lambda'')\ a''_Y = \frac{1.02}{0.17} = 6.0 \text{ dm}^3 \text{ g}^{-1} \text{ cm}^{-1}$$

Hence

$$(1)\ 0.47 = 2.25\, c_X + c_Y$$

$$(2)\ 0.54 = 0.25\, c_X + 6.0\, c_Y$$

This pair of simultaneous equations can be solved very simply. First eliminate c_Y by multiplying the first equation (1) by 3 and subtracting (2) from the result. This gives $c_X = 0.134$ g dm^{-3}. Calculation of c_Y then gives 0.0844 g dm^{-3}.

The final result (to two significant figures in keeping with the experimental data) is:

$$c_X = 0.13 \text{ g dm}^{-3}$$

$$c_Y = 0.084 \text{ g dm}^{-3}$$

Π Do you fully understand why 271 nm was chosen instead of 263 nm as one of the most suitable wavelengths for measurement?

The principal reason is that at 271 nm a peak of *o*-xylene lies at a trough of *p*-xylene thus avoiding measurements on the sloping side of an absorption band which always reduces precision (Section 2.3). But you should also note that at 275 nm the absorption coefficient of *p*-xylene (6.0 dm^3 g^{-1} cm^{-1}) is much larger than that of *o*-xylene (0.25 dm^3 g^{-1} cm^{-1}) so that this system approximates to Case B (Fig. 3.2a), and as a result the precision of analysis, particularly for the *p*-xylene component, is quite high.

The above example is commonly used in teaching laboratories in order to demonstrate the principle of the procedure which is of general application to binary mixtures, for example with acetylsalicylic acid (aspirin) and salicylic acid.

3.2.2 Principles of the Analysis of a Multicomponent System

The above analysis can, in principle, be extended to 3 or more components in a mixture of absorbing species.

Π Can you suggest what conditions must hold? For a 3 component system, how would you select the wavelengths to make the measurements?

The conditions necessary for the above type of analysis are again:

(*a*) the absorbances of the components in the mixture need to be additive, and,

(*b*) each component should obey Beer's Law at all the wavelengths chosen for measurement.

The number of wavelengths used for the analysis must at least equal the number of components, ie 3 wavelengths for 3 components, n wavelengths for n components. However, the feasibility of a particular multicomponent analysis depends on the specificity of the spectra; for a three component system each of the components should show strong absorption in a region where the other two components show small or zero absorption.

When these conditions hold we set up three equations and solve them in the standard manner previously used for the binary mixture

$$\text{at } \lambda' \qquad A'_M = a'_X c_X l + a'_Y c_Y l + a'_Z c_Z l \tag{3.5}$$

$$\text{at } \lambda'' \qquad A''_M = a''_X c_X l + a''_Y c_Y l + a''_Z c_Z l \tag{3.6}$$

$$\text{at } \lambda''' \qquad A'''_M = a'''_X c_X l + a'''_Y c_Y l + a'''_Z c_Z l \tag{3.7}$$

Many three component analysis have been reported and a few involving four components, but the analysis of more complex mixtures is usually not feasible in the uv/visible. Such mixtures require either a number of analytical techniques specific to individual components or the mixture has to be separated by chromatographic techniques (eg hplc) and subsequently subjected to measurement by uv/visible spectrometry.

For example the analysis of products containing a mixture of B vitamins [say thiamin (B_1), riboflavin (B_2), pyridoxin (B_6) and nicotinamide (niacin)] usually involves the application of quantitative hplc. However, two of the components can be separately determined by fluorescence spectrometry (B_1 and B_2), whilst computer-assisted curve-fitting techniques are being developed for direct uv/visible analysis of the mixture. The method being developed depends on the change of the spectra of certain of the components with pH in order to achieve distinct absorption curves.

When a particular multicomponent analysis has been shown to be feasible and is in frequent use, it is possible to simplify the calculations involved by the use of matrix methods. However, these methods are beyond the scope of the present text and you are recommended to consult other sources to obtain details of this approach.

3.2.3 Curve-fitting Techniques

The precision of the analysis of a multicomponent system by uv/visible spectrometry can be improved by fitting the spectrum as a whole using least squares criteria, rather than simply fitting it at *n* wavelengths for *n* components (which is the basis of the methods described in Section 3.2.1 and 3.2.2). In practice this involves collecting absorbance values at close wavelength intervals over the whole of the spectral range of interest. With modern microprocessor-equipped instruments this is readily achieved and the information is stored in digital form. However, the success of the analysis depends on the very high performance of modern spectrometers, particularly with respect to calibration checks, base-line checks, sensitivity and signal-to-noise ratio.

Again the details are beyond the scope of this introductory course and if you are interested reference texts should be consulted.

3.3 ADDITIVE AND NON-ADDITIVE SYSTEMS, PURITY INDICES

The binary and multicomponent analyses detailed in Section 3.2 depend on a knowledge of the absorption characteristics of all the absorbing components in the mixture. It was also assumed that the substances show additive absorbances and obey Beer's Law in the mixture. In this short section we will look first at the problem when non-additivity occurs and then at the methods of data analysis which allow us to use uv/visible spectrometry to assess purity and to analyse the system even when there is background absorption of unknown origin.

3.3.1 Tests for Additivity (Binary Mixtures)

The simplest test is to measure the absorption curve of a synthetic mixture and to compare it with the predicted mixture curve assuming additive behaviour. Figures 3.3a and 3.3b compare this approach for mixtures of two components (blue and yellow dyes) which are known to display additive and non-additive behaviour in different solvent systems. In Fig. 3.3a the absorption curve falls exactly upon the points expected by adding the individual absorbances for the two dyes at any wavelength. But in Fig. 3.3b the observed absorption curve for the mixture does not coincide with the calculated values.

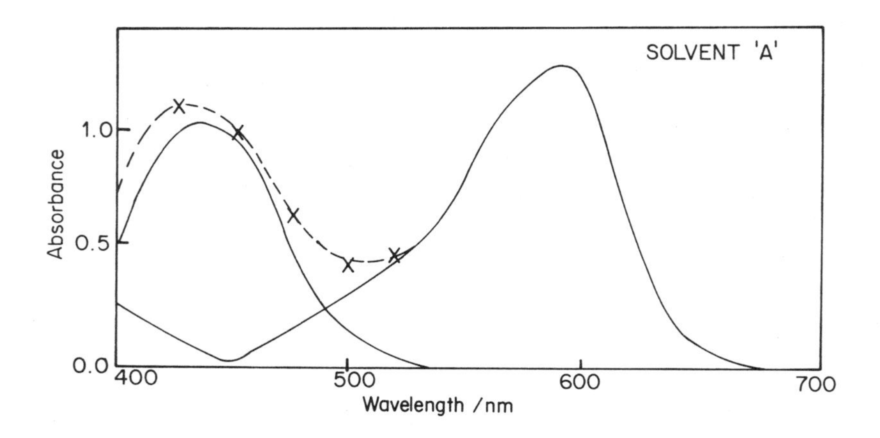

Fig. 3.3a. *Binary Dye Mixture (Yellow and Blue) showing additive behaviour*

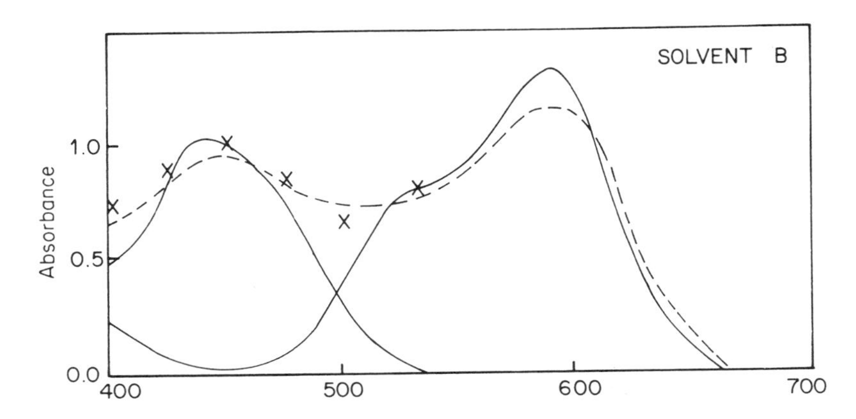

Fig. 3.3b. *Binary Dye Mixture (Yellow and Blue) showing non-additive behaviour*

Non-additivity arises from chemical interaction between the components, which can be influenced by the solvent used. The problem presented as SAQ 3.3a illustrates a binary analysis in which solvent dependency of additivity is illustrated.

SAQ 3.3a

A certain dye A has the same absorption spectrum in aqueous solution containing either 10% pyridine or 10% ethanol. Dye B behaves in the same way. A mixture of the two dyes, however, does not give the same absorption spectrum in the two solvents. Determine from the data below the composition of the mixture, showing how the more satisfactory data are selected. ⟶

SAQ 3.3a (cont.)

	450 nm	550 nm	650 nm
Dye A (1 g dm^{-3})	1.75	0.68	0.21
Dye B (1 g dm^{-3})	0.11	0.23	0.33
Mixture in ethanol-water	2.07	1.75	1.30
Mixture in pyridine-water	2.08	1.37	1.20

(Figures given in the tables are absorbances in a 10 mm cell at the wavelengths indicated)

To help you start, treat the problem as one involving a binary mixture, choose the two most appropriate wavelengths for the analysis, and then check that the result fits at the third wavelength. You will have to decide which solvent system to choose for the calculation - only one solvent gives satisfactory results.

SAQ 3.3a

If a change of solvent does not eliminate non-additivity, the only recourse left to the analyst is to evaluate empirical corrections for the absorption coefficients of the components in the mixture based on studies of synthetic mixtures covering the range of concentrations or properties likely to be met in practice.

3.3.2 Impurity and Background Absorption

Many natural products which are analysed by uv/visible spectrometry contain impurities which contribute some background absorption or scattering over the wavelength region used in analysis. The classic example is vitamin A in fish liver oils, where the absorption band at 328 nm is distorted due to impurity or background absorption tailing from the low uv end of the spectrum.

A correction procedure for the analysis of this situation was developed by Morton and Stubbs and is known by their names. In it the assumption is made that in the region of measurement the background absorption varies linearly with wavelength.

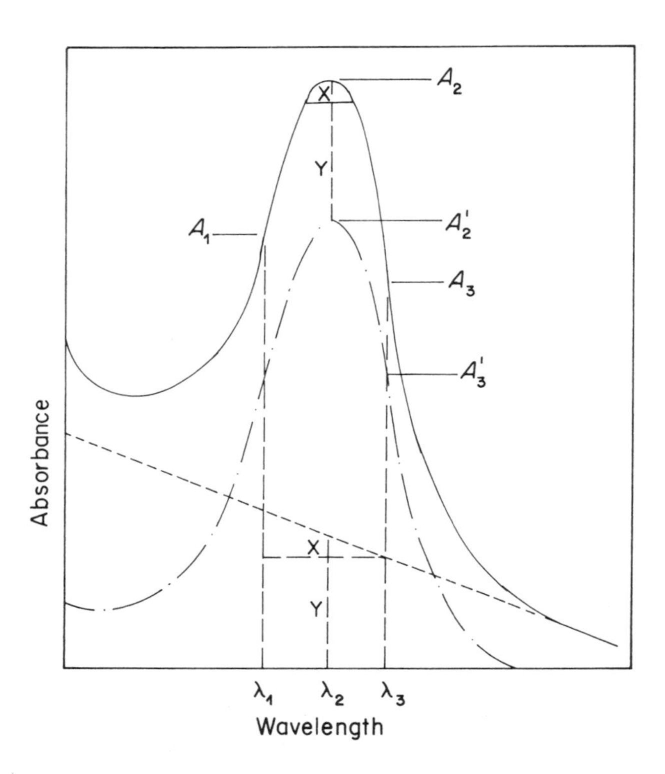

Fig. 3.3c. *The Morton and Stubbs baseline correction. A_1, A_2 and A_3 are the absorbances at three points on the observed absorption spectrum (—). A'_2 is the absorbance at λ_2 on the true absorption spectrum (-.-.-) after correcting for the background absorption (—) which is assumed to be linear between λ_1 and λ_3. Redrawn from Morton and Stubbs*

In terms of quantities indicated on the diagram in Fig. 3.3c we have the corrected absorbance

$$A'_2 = A_2 - (x + y) \tag{3.8}$$

where $x + y$ is the assumed background absorption.

The wavelengths λ_1 and λ_3 are chosen at appropriate positions on the absorption curve so that in *pure* Vitamin A, $\varepsilon_1 = \varepsilon_3$ and $\varepsilon_2/\varepsilon_3 = R$ (a known ratio).

Using the property of similar triangles it is then possible to show that:

$$x = (A_1 - A_3)(\lambda_3 - \lambda_2)/(\lambda_3 - \lambda_1) \quad (3.9)$$

and

$$y = [RA_3 - (A_2 - x)]/(R - 1) \quad (3.10)$$

Hence the baseline is now defined.

You will have some feeling for the magnitude of the corrections by taking some typical results quoted by Morton and Stubbs for a cod liver oil extract.

Π Extraction of Vitamin A from fish oil followed by absorbance measurements of the cyclohexane extract gave the following results.

$$\lambda_1 = 313 \text{ nm}, \quad A_1 = 0.640$$

$$\lambda_2 = 328 \text{ nm}, \quad A_2 = 0.712$$

$$\lambda_3 = 338.5 \text{ nm}, \quad A_3 = 0.620$$

Given that $R = \varepsilon_2/\varepsilon_3 = 1.67$, show that the correction for the absorbance of the background amounts to 17.6% of the measured values at 328 nm.

Substitution in the three equations above should give

$$x = 0.008 \quad \text{and } y = 0.117$$

hence

$$A_2' = 0.587$$

So the correction at 328 nm is

$$\frac{(0.712 - 0.587)}{0.712} = \frac{0.125}{0.712} = 17.6\%$$

The presence of background absorption in the above example can be deduced by the deviation of the A_2/A_3 or A_2/A_1 ratio from the expected value R, and such ratios are useful measures of purity as discussed below. However, various mathematical methods have been used to handle situations where the background absorption is either flat or curved and these methods are being incorporated in curve-fitting techniques developed as part of the armoury of microprocessor linked instrumentation.

3.3.3 Purity Checks and Purity Indices

Uv/visible spectrometry is widely used in the pharmaceutical industry as a means of quality control, mainly to check tablet dissolution rates and the concentration of uv absorbing components. To check for impurity absorption it is necessary to have absorption data for the pure compound, and log A plots against wavelength are useful for this purpose (see Part 4). However the possibility of using absorbance ratios as checks of purity has already been mentioned.

Π Suppose you were developing a method for the purity check of a synthesised Vitamin A material, based on the measurements of absorbance at λ_3 (338.5 nm) and λ_2 (328 nm). What instrumental factors would affect the measured absorbance ratio at these two wavelengths?

First of all there would be the accuracy of *wavelength calibration*, which would need to be checked routinely.

Secondly measurements at 338.5 nm are on the sloping side of the absorption band and are subject to variations with slit-width, which would need to be specified and kept constant.

For the second reason purity indices are usually based on the measurement of absorbance ratios of peak absorption values in the spectrum of the compound being tested, or at least peak to trough absorbance values.

3.4 APPLICATIONS OF SPECTROMETRIC ANALYSIS

Π To remind yourself of the variety of substances dealt with, list the applications which have been mentioned in this part?

Iron in water

Glucose in blood

Isomer content of xylene

Vitamin A in fish liver oil

Vitamin B mixtures

Dye mixtures in solution

Iron in reagent chemicals

These represent just a few examples drawn from the wide areas of application mentioned in Section 1.4.

Π Can you, for each of the above, suggest the industries or institutions which would be most concerned with the above types of analysis?

The obvious answers are in order, the water supply industry, hospitals, the petrochemical industry, the pharmaceutical industry (twice), the textile dyeing industry and finally the standardising laboratories of a reagent supply company. However, concern for water quality arises in many different situations, and glucose content is of concern to the individual diabetic patient or his doctor in his surgery. Thus chemical analytical techniques have to be applicable in a wide variety of environments, not necessarily all equipped with the best instrumentation for the analysis in question. So simple procedures are as important as those using advanced, computerised instruments.

Thus, although atomic absorption spectrometry may be the preferred technique for iron in water, the spectrometric method, or even a simple colorimetric analysis, based on the use of phenanthroline or TPTZ reagents has to be available for those who do not have access to an atomic absorption spectrometer. Similarly the combination of absorption and fluorescence spectroscopy may have to be used for Vitamin B mixture analysis where an hplc method is not available.

Summary

Many substances to be determined by chemical procedures are mixtures with the components having similar or related characteristics. Uv/visible spectrometry is capable of dealing with binary mixtures of substances with overlapping spectra by making use of the Beer-Lambert Law at two different wavelengths. This procedure can be extended to multicomponent systems, so long as each substance has a clearly identified maximum absorption.

Objectives

You should now:

- be aware that there are numerous sources available of very well defined spectrometric procedures;
- understand the need for precision and accuracy in uv/visible measurements;
- be able to follow the steps taken in a quantitative ultraviolet measurement;
- have a knowledge of the mathematical procedures used in obtaining quantitative results from a binary mixture;
- be able to carry out calculations using absorption values from binary mixtures.

4. Qualitative Analysis and Structural Relationships

4.1 IDENTIFICATION BY ABSORBANCE PLOTS

In Part 1 of the present Unit we reviewed the use of colour tests in the identification of chemical species, mention being made of the characteristic purple colour of aqueous potassium permanganate solutions familiar to most students of chemistry. But if we relied only on the colour as perceived by the eye we could on occasions be mislead. Any particular solution colour can actually arise from a wide range of different compounds – think of all the solutions which might be described as yellow.

Fig. 4.1a shows the visible absorption spectrum of a solution of potassium permanganate along with the absorption curve of a solution of a soluble azo dye which appears to the eye to have the same purple colour as the permanganate solution.

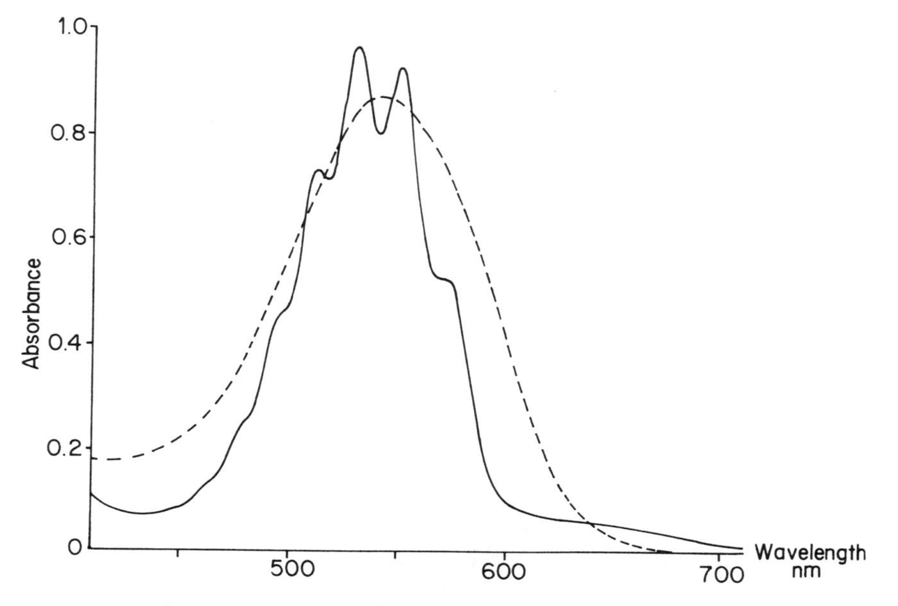

Fig. 4.1a. *Absorption Spectra of two substances with identical colours*

In viewing solutions of these two substances the eye registers identical colours because both species have an absorption band centred on the green region near 530 nm; the absorptions are also of approximately the same intensity.

Although the colours appear identical the shapes of the absorption curves, as recorded by the spectrometer, are obviously different, and in using absorption curves in the uv and visible for purposes of identification or qualitative analysis it is important not only to specify the position and intensity of the absorption bands but also to indicate in some way the shape of the absorption curve.

As we have seen previously in this Unit the position of an absorption band in the uv/visible is usually defined by quoting the wavelength (λ) in nm units, although occasionally wavenumbers ($\bar{\nu}$) in cm^{-1} or even frequencies (ν) in Hertz are used. The relationships between these has already been given, and you may recall that it is:

$$\lambda\nu = c \therefore \nu = c/\lambda \text{ and } \bar{\nu} = 1/\lambda$$

The velocity of light (c) being taken as 3 x 10^8 m s^{-1}.

You will also know that the intensity of any particular absorption band in the uv/visible is usually quoted in terms of the molar absorptivity ε, as defined in the Beer-Lambert expression:

$$A = \varepsilon c l$$

Here concentration terms (c) need to be expressed in mole terms (usually mol dm^{-3}) and the cell path length in mm or cm. Where the molecular mass of a substance is unknown or the purity of a product is unspecified (e.g. with dyes and many natural products), it is common practice to use concentrations (c') expressed in g per 100 ml (% wt/vol) units, when the absorptivity is represented by the symbol $E_{1\%}^{1cm,}$ and the Beer-Lambert expression becomes:

$$A = E_{1\%}^{1cm,} c' l \text{ with } l \text{ in } cm \text{ units}$$

Π Use the information in Fig. 4.1a to evaluate the intensity of the permanganate and azo dye absorption in the following units.

	Position Unit	Intensity Unit
(a)	λ_{max}/nm	$E_{1\%}^{1cm}$
(b)	$\bar{\nu}_{max}$/cm^{-1}	ε_{max}/dm^3 mol^{-1} cm^{-1}
(c)	ν_{max}/Hz	ε_{max}/m^2 mol^{-1}

You should have obtained the following values, although the actual results will depend upon the figures obtained for the absorbances at λ_{max}.

	$KMnO_4$	Azo Dye
λ_{max}/nm	533	542
ν_{max}/cm^{-1}	18760	18450
ν_{max}/Hz	5.62×10^{14}	5.53×10^{14}
$E_{1\%}^{1cm,}$	640	435
ε_{max}/dm^{3}mol^{-1} cm^{-1}	1.01×10^4	1.88×10^4
ε_{max}/m^2 mol^{-1}	1.01×10^3	1.88×10^3

The azo dye has an unstructured absorption curve with a single peak, whereas the permanganate absorption shows some structure. This may be indicated by quoting absorption characteristics of $KMnO_4$ in the form

$$\lambda/\text{nm}\ (496),\quad 514,\quad 533,\quad 553,\quad (570)$$

which indicates a triple peak with shoulders in the absorption at 496 and 570 nm.

The wavenumber values in cm^{-1} are obtained by using $\bar{\nu} = 10^7/\lambda$ with λ in nm.

The frequency values in H_z are similarly obtained from $\nu = 3 \times 10^{17}/\nu$.

Note that the final unit for ε_{max} is in SI units derived using c in mol m^{-3} and l in m.

Thus the two purple compounds in Fig. 4.1a are readily distinguished because of the characteristic structured appearance of the $KMnO_4$ spectrum. The shoulders and side bands arise due to the spectrometer separating the vibrational structure of the electronic transition in the MnO_4^- species. Such structure is relatively unusual in spectra of solutions in the uv/visible, although of course the detailed vibrational pattern in an infrared spectrum leads to the description of such spectra as *chemical fingerprints.*

Another distinguishing characteristic is the width of the absorption band, but this is rarely quoted in tables of absorption characteristics. The $KMnO_4$ spectrum has a narrower or sharper absorption band than the dye.

Π Can you suggest how bandwidths might be measured?

From what has been covered in earlier sections you should be aware that it is by quoting bandwidth in nm at half-peak height.

It should be apparent to you from this introductory discussion that to use uv/visible absorption data for qualitative analysis we require more than just peak position and intensity data. Various means have been devised for distinguishing materials which have similar absorption characteristics varying from curve shape indices to the use of logarithm of absorbance (log A) plots. The changes in these quantities or plots when the solution conditions are altered or the chemical species undergoes reaction with an identifying reagent are also useful. Note that the absorbance ratio method of checking purity mentioned in Part 3 is a simple curve shape parameter.

As you might expect there is a relationship between absorption characteristics and chemical structure. The final part of the section will examine how changes in uv/visible absorption are associated with structure in organic compounds and how these changes can be used to investigate solution reactions and equilibria.

4.2 PRESENTATION OF ABSORPTION DATA IN THE UV AND VISIBLE

As indicated in the introductory Section 4.1 above the simplest method of characterising a material which has a single absorption band in the uv or visible is to quote the position and intensity of the absorption in suitable units. Most of the standard texts on the uv/visible spectra of organic compounds utilise tables of λ_{max} (in nm) and ε_{max} (in dm^{-3} mol^{-1} cm^{-1}). Fig. 4.2a illustrates a typical table of values, remember that these values of ε should be divided by 10 to give the corresponding m^2 mol^{-1} values.

Type	Compound	λ_{max}/nm	ε_{max}
Linear dienes	1,3-Butadiene	217	21 000
	Isoprene	220	23 000
	2,4-Hexadiene	227	23 000
Semicyclic dienes	β-Phellandrene	231	9 100
	Cyclohex-1-enylethylene	230	8 500
	Menthadiene	235	10 700
Cyclic dienes	Cyclopentadiene	238	3 400
	Cyclohex-1,3-diene	256	8 000
	Cyclohepta-1,3-diene	248	7 500
Polycyclic dienes	L-Pimeric acid	273	7 100
	Ergosterol	280	13 500
	7-Dehydrocholesterol	280	11 400
	7-Dehydrocholestene	280	12 700
	Cholesta-3,5-diene	235	23 000
	Cholestadienol-C	248	17 800
	Ergosterol-D	242	21 400
	Abietic acid	238	16 100

Fig. 4.2a. *Ultraviolet Absorption Spectra of Dienes*

(*Note that a mixture of IUPAC nomenclature and trivial chemical names is used in this table.*)

This type of information is only a general guide to the absorption characteristics of a group of compounds since the precise values of the λ_{max} and ε_{max} will depend on instrumental characteristics, solvent and solution conditions and material purity.

Full absorption spectra are generally more useful although with all the alternative scales possible for position and intensity, methods of presentation of data are still the subject of controversy.

In some libraries the spectra are plotted using log ε as the vertical intensity axis and cm^{-1} (wavenumber) as the horizontal axis to indicate position (with corresponding nm wavelength units as a parallel scale). The position and intensity of all major peaks in the spectrum may also be presented in tabular form. Details of compound purity, instrument used, nature of solution, and conditions of measurement should all be listed since all of these are factors which can influence the absorption curve. The use of such collections of spectra can be helpful when attempting to apply uv/visible spectroscopy to the analysis of compounds for which standard spectra of your own are not available. However, when employing published spectra you must try to avoid confusion because of the different definitions, symbols and methods of presentation used in the various commercial publications and literature sources. The most reliable analyses will be the ones which utilise your own data since you will be aware of sample purity and solution conditions, and will be making comparisons using spectra recorded on the same instrument (hopefully).

SAQ 4.2a

Fig. 4.2b is an absorption spectrum of potassium permanganate in which the four major peaks have been numbered. Values for λ_{max} and ε_{max} for the strongest peak are given. Measurements were made using a 1 cm cell.

Complete the table below to show the λ_{max} and ε_{max} values of the three major bands

	λ_{max}/nm	ε_{max}/$m^2 mol^{-1}$
Band 1	533	1.01×10^3
Band 2		
Band 3		
Band 4		

⟶

SAQ 4.2a (cont.)

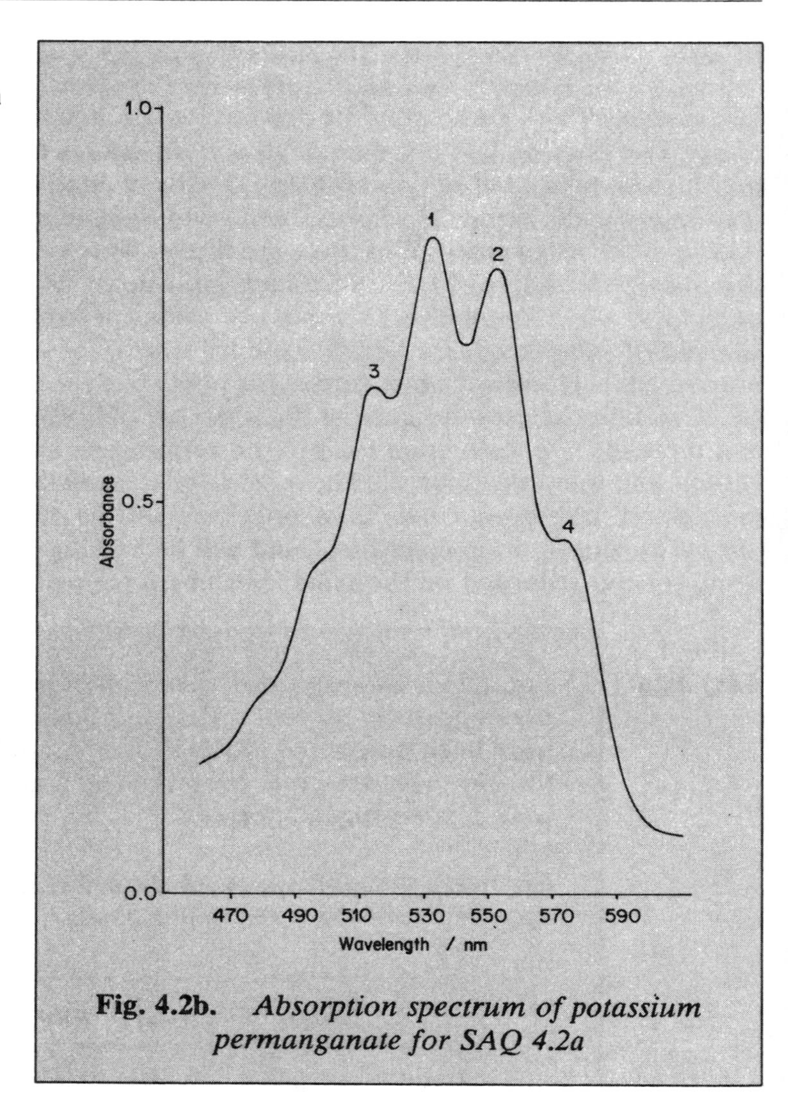

Fig. 4.2b. *Absorption spectrum of potassium permanganate for SAQ 4.2a*

SAQ 4.2a

4.3 LOG *A* CURVES AND CURVE SHAPE PARAMETERS

The shape of a measured absorption curve in the uv/visible depends on the variation of absorbance with wavelength, which in turn depends on the variation of ε and λ. Expressed in terms of the Beer-Lambert law we can write:

$$A_{\lambda} = \varepsilon_{\lambda} lc \qquad \text{and taking logs;}$$

$$\log A_{\lambda} = \log \varepsilon_{\lambda} + \log lc$$

The shape of the log A_{λ} curve as distinct from the A_{λ} curve is independent of path length and concentration. This is illustrated in Fig. 4.3a in which four different solutions of potassium permanganate have been studied using the two techniques.

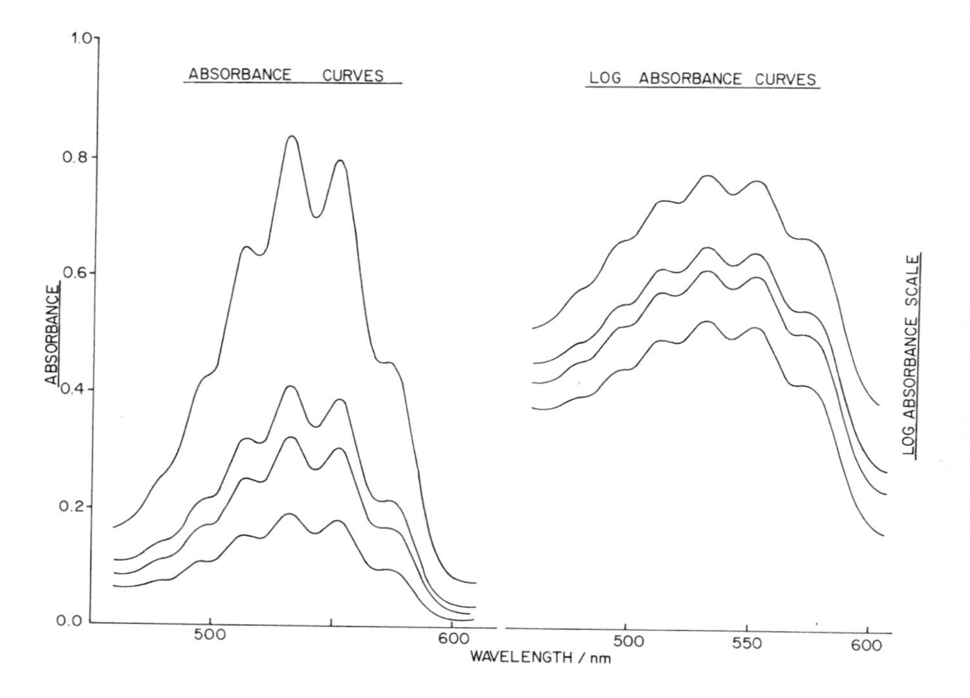

Fig. 4.3a. *Comparison of Absorbance and Log Absorbance Spectra*

If the log A_λ curve of a known standard compound is prepared on a transparent sheet it is possible to check for identity or purity simply by overlaying the standard curve on that of a sample being tested. Coincidence of the curve shapes is a very good test of identity. This procedure has been extensively used in the dye and pigment industry for qualitative analysis and purity checks. It is of course applicable to the absorption spectra of all classes of compounds, and it is recommended that where absorption spectroscopy is being used as a quality control device the absorption data of the standards should be held in log A_λ form. Most modern recording uv/visible spectrometers have a log A_λ facility.

Let us finally consider the relationship of the log A_λ method to the peak ratio method of checking purity; the latter method was mentioned in Part 3.

Suppose a solution of a standard compound has two peaks with absorbance values of $A_{\lambda_1} = 0.2$ and $A_{\lambda_2} = 0.4$. For this solution *the peak ratio*

$$A_{\lambda_2}/A_{\lambda_1} = 0.4/0.2 = 2 : 1$$

At double the concentration the peak ratio is the same

$$A'_{\lambda_2}/A'_{\lambda_1} = 0.8/0.4 = 2 : 1$$

with a log A_λ facility we would measure *the difference* in peak heights

$$\text{ie } \log A_{\lambda_2} - \log A_{\lambda_1} = \log 2 = 0.301$$

which would *be the same* for both solutions.

4.4 SPECTRA-STRUCTURE CORRELATIONS IN ORGANIC CHEMISTRY

Attempts to find relationships between colour and chemical constitution have stimulated the endeavours and imaginations of chemists for over 100 years. However, it was only with the introduction of quantum concepts and theoretical calculations based on wave mechanics that the complex nature of these relationships were gradually appreciated. Even today molecular orbital calculation methods are still being developed to explain the origin of colour in chemical species such as dyes and transition metal complexes.

In this section we shall consider some of the empirical relationships which have been useful in elucidating the structure of simple organic compounds which absorb in the accessible uv/visible (180-700 nm) region of the spectrum.

The work of the early chemists on the colour and constitution of organic compounds introduced terms such as chromophore (used for a chemical grouping which had the property of conferring colour on a substance) and auxochrome (which relates to a group not capable of producing colour on its own but possessing the power to modify or enhance the colour produced by the chromophore). We now use the term chromophore to indicate a group capable of producing absorption in the 200-800 nm range, with the auxochrome shifting the absorption to longer wavelengths (known as red or bathochromic shift).

Some typical chromophoric groupings are illustrated in Fig. 4.4a, along with the absorption positions for simple compounds containing these chromophoric groups.

Group	Compound	λ/nm
$>C=C<$	$CH_2=CH_2$	180
	C_6H_6	255
$>C=O$	$(CH_3)_2CO$	277
$-N=N-$	$CH_3-N=N-CH_3$	347
$>C=S$	$(CH_3)_2C=S$	400
$-N=O$	$C_4H_9N=O$	665

Fig. 4.4a. *Chromophoric groups*

Fig. 4.4a indicates the position only of the longest wavelength absorption band, but gives no indication of the intensity of absorption (the ε_{max} values vary from 10 to 10^5). For example acetone $(CH_3)_2CO$ shows three absorptions of differing intensity which have been assigned to three different electronic transitions.

Position	ε_{max}/m^2 mol^{-1}
180 nm	10^4
205 nm	10^3
277 nm	10^1

The effect of an auxochrome (-NH_2 group) on characteristic absorptions can be seen from Fig. 4.4b.

Compound	λ_{max}/nm	ε_{max}/m^2mol^{-1}
$CH_2{=}CH_2$	180	10^3
$CH_2{=}CHNH_2$	220	10^4
(benzene)	255	23
(benzene)- NH_2	280	143

Fig. 4.4b. *Effect of the auxochrome* $-NH_2$, *on* λ_{max} *and* ε_{max} *of* $>C{=}C<$ *and* (benzene)

In both cases introduction of the auxochrome leads to a higher absorption at longer wavelength.

Combining chromophores or extending conjugation can also significantly affect position and intensity as shown in Fig. 4.4c.

Position/ Intensity	λ_{max}/nm	ε_{max}/m^2mol^{-1}	λ_{max}/nm	ε_{max}/ m^2mol^{-1}
$CH_2{=}CH_2$	180	10^3	–	–
$CH_2{=}CH{-}CH{=}CH_2$	217	2.1×10^3	–	–
$H_2C{=}O$	180	10^3	273	1.2
$CH_2{=}CH{-}CH{=}O$	217	1.6×10^3	321	2.0

Fig. 4.4c. *Effect of conjugation on* λ_{max} *and* ε_{max}

How can we explain these effects? With conjugation the greater the number of π electrons in the system the less energy is required to promote one of them into the excited state, and hence the λ_{max} increases. The delocalisation effect of conjugation reduces the gap between the ground state and the excited state. The value of ε_{max} tends to increase in line with the number of electrons in the conjugated system.

The auxochromic effect depends on the ability of the chemical group to donate electrons into the conjugated system. This has been most studied with aromatic systems and the spectral shifts of monosubstituted aromatic compounds have been correlated with electron donating power. The electron-donating properties of some common substituents decrease in the order

$$O^- > NHCH_3 > NH_2, OH > Cl > CH_3 > NH_3^+, H$$

In this list the significant effect of protonating the NH_2 group should be noted, the proton binds the non-bonding electrons on the nitrogen of the amino group and prevents them from interacting with the benzene π electron system.

SAQ 4.4a

Assign from list 2 below, the appropriate values of λ_{max} and ε_{max} for each system given in list 1. Comment on the shifts observed relative to benzene λ_{max} = 255 nm; ε_{max} = 23 $m^2 mol^{-1}$ in terms of the electronic properties of the auxochromes.

List 1 Absorbing systems

A aqueous solution of aniline

B solution A with added 0.01 M HCl

C aqueous solution of phenol

D solution C with added 0.01 M NaOH

⟶

SAQ 4.4a (cont.)

List 2 Absorption Characteristics

	λ_{max}/nm	ε_{max}/m^{2}mol^{-1}
E	255	20
F	270	140
G	280	130
H	290	230

Further detailed discussion of the structural influences of chromophores and auxochromes on the absorption of chemical species is outside the scope of the present treatment – if you are interested in extending your knowledge of these effects you should study the texts listed in the Bibliography. However, we will study the use of Woodward's rules for predicting the absorption of diene systems as just one example of a quantitative although empirical correlation between uv absorption and chemical structure.

4.5 WOODWARD RULES FOR DIENE ABSORPTION

A number of important steroids contain polycyclic diene structures and Woodward developed rules relating their uv absorption to structural characteristics. The rules provide values for calculating wavelengths for maximum absorptions based upon a parent diene as a result of a red shift arising from extension of the conjugation and/or the presence of various auxochromes.

Parent acyclic diene	217 nm
Parent heteroannular diene	214 nm
Parent homoannular diene	253 nm
Parent α, β unsaturated carbonyl	222 nm
Increments for	
Double bond extending conjugation	30 nm
Alkyl group or ring residue	5 nm
Exocyclic double bond	5 nm
Polar groups (for each one)	
O(acyl)	0 nm
O(alkyl)	6 nm
S(alkyl)	30 nm
Cl, Br	5 nm
$N(alk)_2$	60 nm
Total gives λ_{max}	

Fig. 4.5a. *Woodward rules for diene absorption (EtOH solutions)*

In order to be able to apply these rules we must be able to identify the types of structures referred to in Fig. 4.5a.

The basic chromophore unit is 1,3-butadiene which is considered the parent acyclic (or non-cyclic) diene.

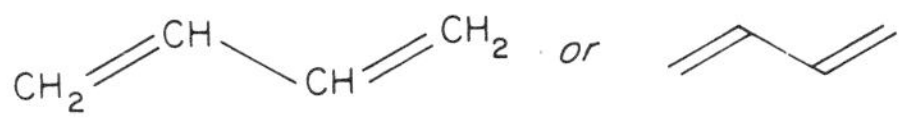

$CH_2{=}CH{-}CH{=}CH_2$ or

If the diene has attached to it saturated alkyl groups, then an additional contribution for each group is added.

eg $CH_3{-}CH{=}C(CH_3){-}CH{=}CH{-}CH_3$ or (a) (b) (c)

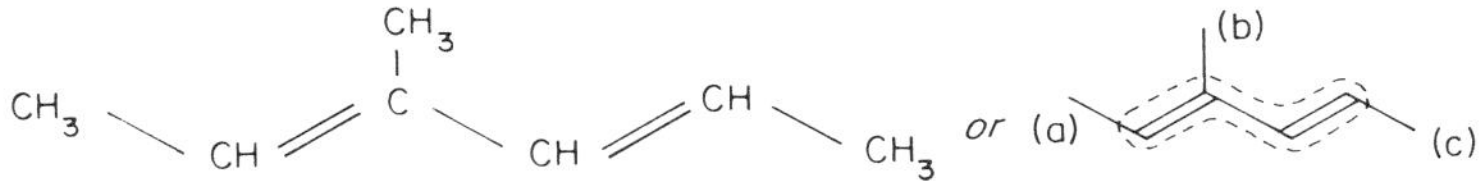

I

This is analysed in terms of the rules as follows;

Base values for I (acyclic structure)		217 nm
For three alkyl or methyl groups (a,b,c) add 3 × 5	=	+15 nm
giving predicted λ_{max}	=	232 nm
compared with observed λ_{max}	=	231 nm

If the diene system is contained in a single ring it is termed *homoannular*, whereas if it is spread over two rings it is said to be *heteroannular*.

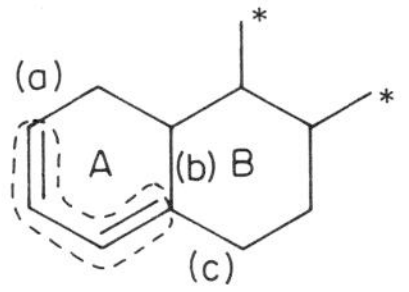

(a) A (b) B (c) II homoannular

eg For compound II we have a base value		253 nm
The ring residues (a,b,c) are all attached to the diene system; add 3 × 5	=	+15 nm
The lower double bond in A is attached to, but outside (or exocyclic) to ring B		+15 nm
Predicted λ_{max}	=	273 nm
Observed λ_{max}	=	275 nm

Note that the groups marked with * do not contribute as they are not directly attached to the diene system.

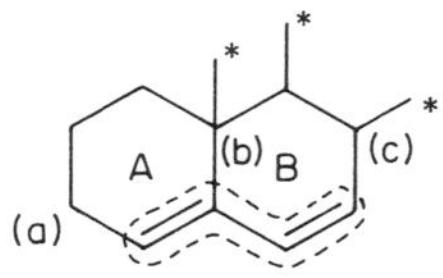

(a) A (b) B (c) * * *

III heteroannular

For compound III we have a base value		214 nm
The ring residues (a,b,c); add 3 × 5	=	+15 nm
The double band in A is *exocylic* to B	=	+ 5 nm
Predicted λ_{max}	=	234 nm
Observed λ_{max}	=	235 nm

Also if a ring structure such as IV occurs:

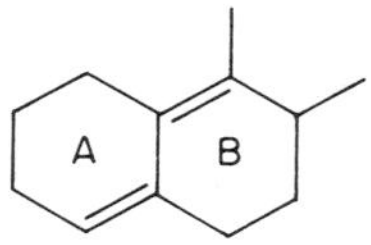

A B IV

then both double bonds are in exocyclic positions hence compound IV would have a 10 nm contribution from this element of its structure.

Π Identify the diene structure and the contributing elements in compound IV and use the Woodward rules to predict λ_{max} in the uv absorption spectrum.

Your calculation should show that:

$$\lambda_{max} = 214 + 25 + 10 = 249 \text{ nm}$$

Note that one of the methyl groups attached to ring B influences the diene absorption, along with the four ring residues attached to the chromophore, giving in total 5 x 5 = 25 nm.

Similar rules apply for α, β-unsaturated carbonyl compounds. The complete tables of rules also include corrections for solvent shifts since the electronic transition in the carbonyl group is sensitive to the polarity of the solvent.

SAQ 4.5a

A compound with the following structure

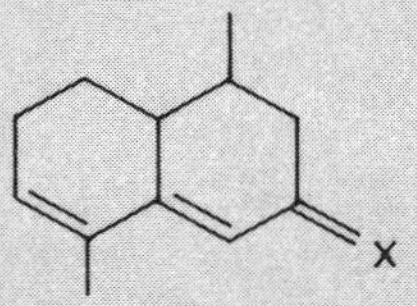

is classed as (*i*) an alkene if X is CH_2,

and (*ii*) an unsaturated ketone if X is O.

For the purposes of applying the Woodward rules to predict the λ_{max} of the uv spectrum in alcohol solution we need to be able to characterise aspects of the structural contributions.

(A) In this context indicate which of the following statements are *true* and which are *false.*

(*i*) When X is CH_2 we have a triene and the diene rules do not apply.

(*ii*) The compound has only *two* exocyclic double bonds.

(*iii*) The compound has two methyl groups both of which contribute +5 nm to the absorption.

(*iv*) When X is an oxygen atom (O) the double bonds of the unsaturation are in the α, β and γ, δ positions.

(B) Using the appropriate rules, show that the predicted λ_{max} (EtOH):

(*i*) for the alkene structure is 269 nm,

(*ii*) for the ketone structure is 277 nm.

SAQ 4.5a

You may wish to try your hand at applying Woodward's rules. The following structures are given with both the observed λ_{max} and the calculated value (which you should obtain without too much difficulty) in parentheses.

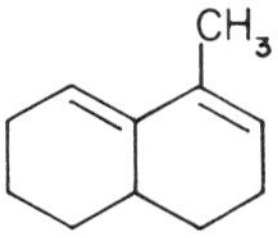

CH_3

238 (239)

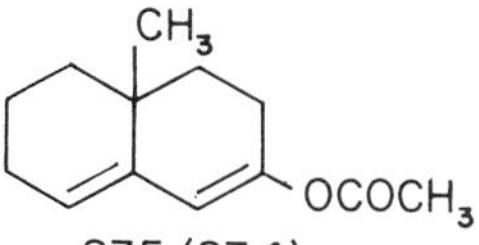

CH_3 $OCOCH_3$

235 (234)

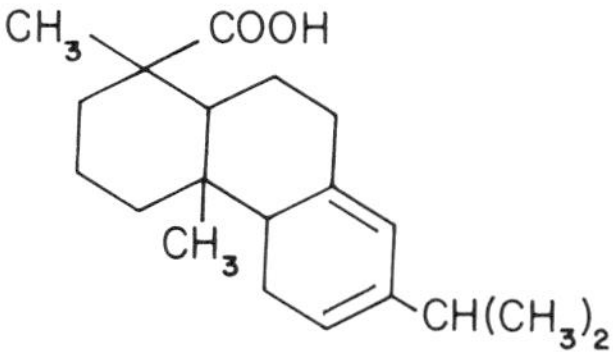

CH_3 COOH CH_3 $CH(CH_3)_2$

273 (278)

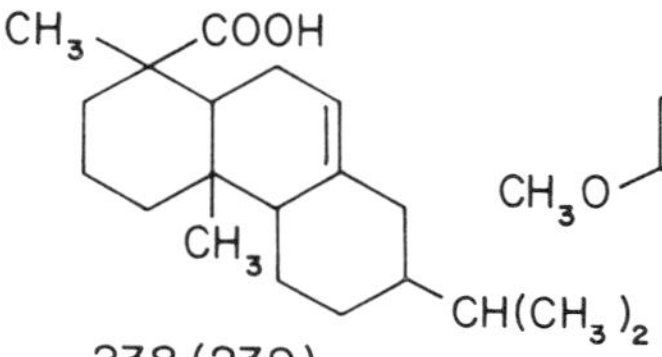

CH_3 COOH CH_3 $CH(CH_3)_2$

238 (239)

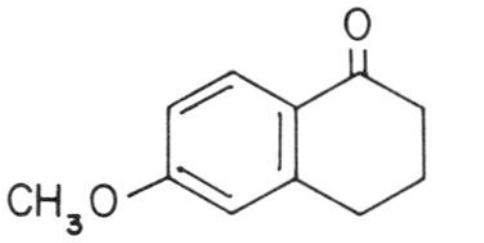

O CH_3O

276 (274)

Cl CH_3 COOET OH O

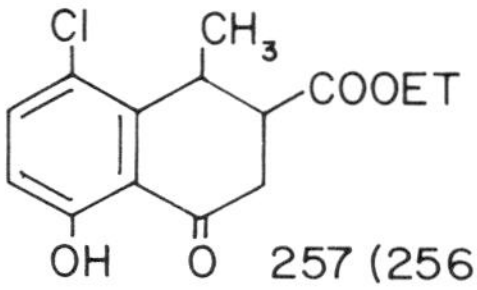

257 (256)

$(CH_3)_2C=CHCOCH_3$

237 (239)

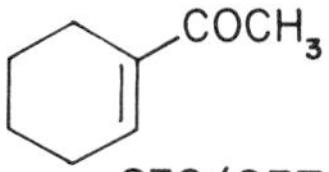

$COCH_3$

232 (237)

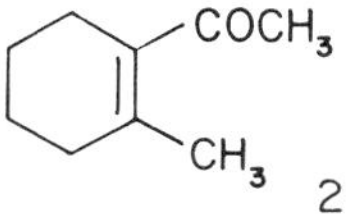

$COCH_3$ CH_3

243 (249)

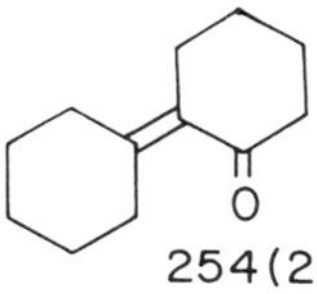

O

254 (254)

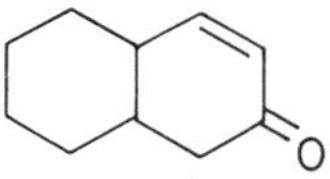

O

225 (227)

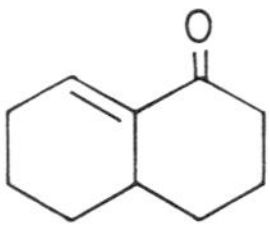

O

241 (242)

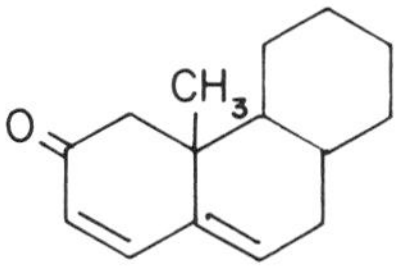

O CH_3

290 (286)

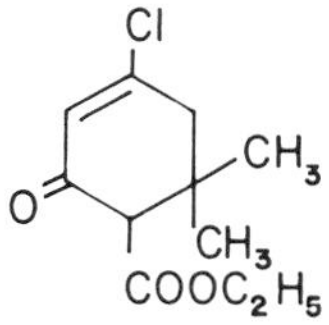

Cl CH_3 O CH_3 $COOC_2H_5$

240 (239)

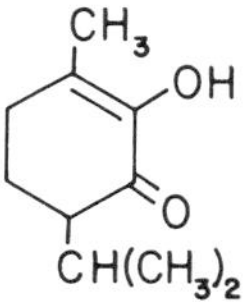

CH_3 OH O $CH(CH_3)_2$

270 (274)

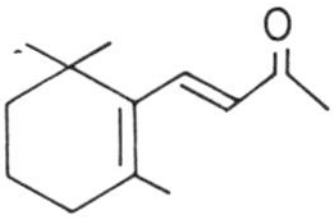

O

294 (299)

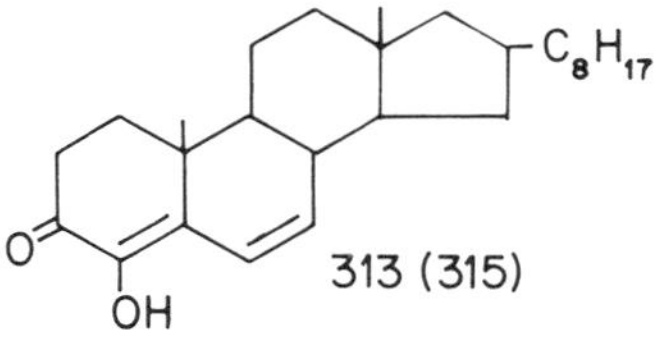

C_8H_{17} O OH

313 (315)

OCH_3 CH_3O O

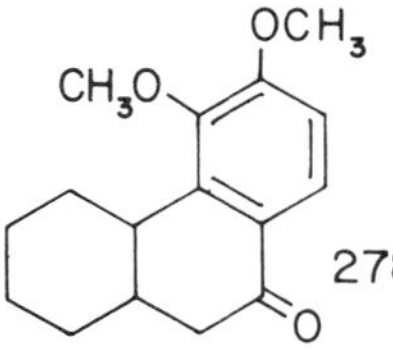

278 (289)

4.6 FOLLOWING REACTIONS AND EQUILIBRIA

Although the use of uv/visible spectroscopy as a means of investigating molecular structure is limited in comparison with ir and nmr spectroscopy, it remains an important tool for the investigation of reactions and equilibria in solution. Any change which affects the electronic properties of the absorbing species in the solution can provide useful means of investigating the change. Some examples of reactions which have been studied by uv/visible spectroscopy include:

Oxidation of unsaturated oils

Methylation of amino groups in dye synthesis

Ligand replacement reactions in complex ions

Cis-trans isomerisation in conjugated aromatic systems

Fast reactions studied by flash photolysis techniques.

Examples from the first and the last of these categories are given below as illustrations of the use of uv/visible spectrometry to study reaction mechanism and reaction kinetics respectively.

The study of chemical equilibria is also an important application of uv/visible spectroscopy. Equilibria which have been studied by uv/visible spectroscopy include:

acid-base equilibria of indicators

chlorophyll-protein complexes

dye-surfactant complexes

dimerisation or aggregation studies

charge-transfer complexes.

The difficult problems in such equilibria studies are:

(*i*) to identify those situations in which a stoichiometric equilibrium between two or more species exist,

(*ii*) to identify the species in equilibrium, and

(*iii*) to devise methods of obtaining the absorption characteristics of the individual species.

In the present section we will restrict ourselves to these problems, although there are many other aspects which are outside the scope of this text.

4.6.1 Reaction Studies Involving a Chemical Change

It has been known for a long time that the thermal oxidation of unsaturated oils involves a free radical mechanism in which peroxides and more complex oxidation products are formed at the double bonds in the oil molecules. The possibility of photosensitised initiation by singlet oxygen has been studied and the extent of monoperoxide formation followed by uv spectroscopy. A typical set of results from the photoxidation of methyl linoleate is illustrated in Fig. 4.6a.

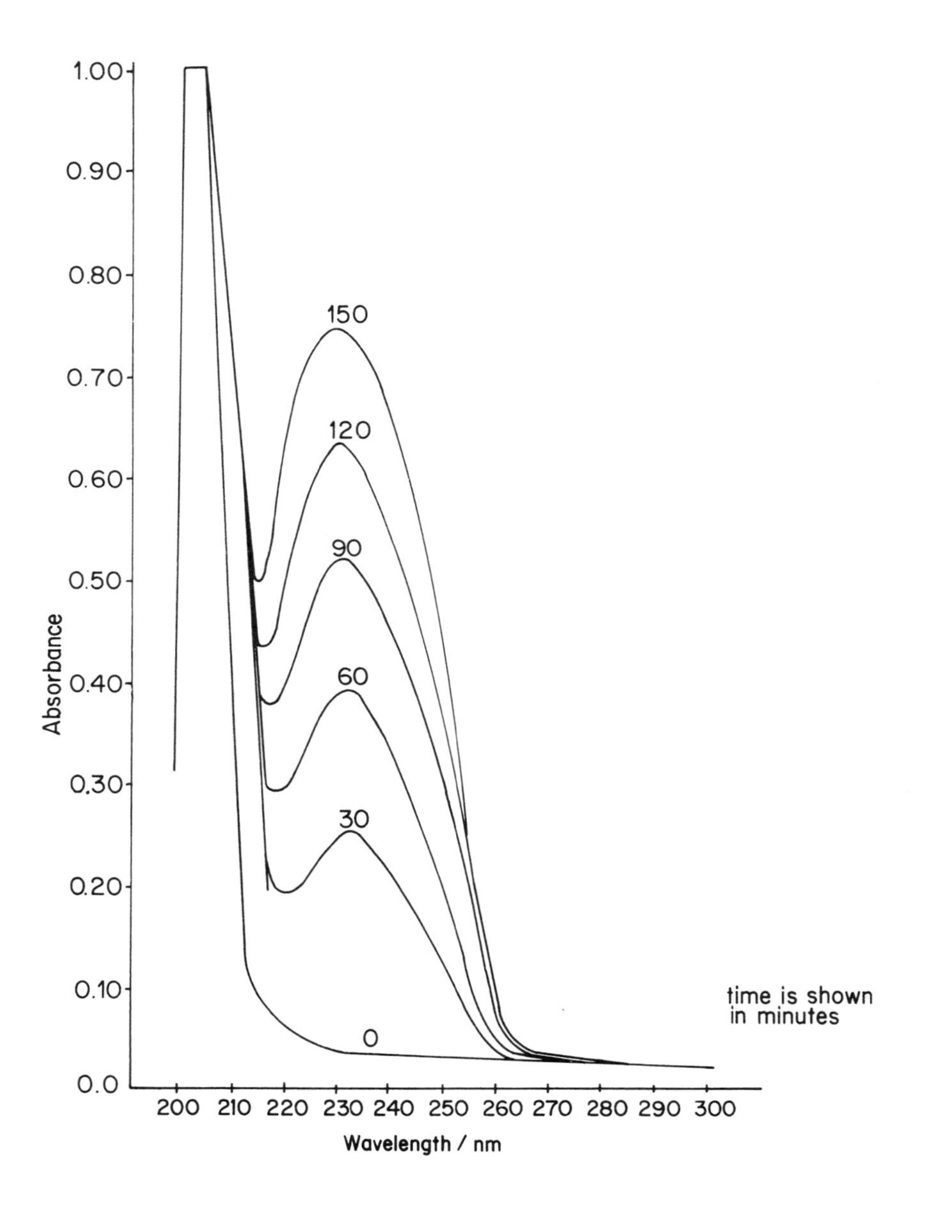

Fig. 4.6a. *Change in the uv spectrum of methyl linoleate/erythrosin (0.04 mol dm^{-3} ester) with irradiation time (diluted 1:100 for measurement)*

It is now known that such a photoxidation produces four possible monohydroperoxides as shown in the reaction scheme noted below.

Linoleate
12 9
$^1O_2^*$
A OOH 12 + OOH 10 D
B OOH 13 + OOH 9 C

Π Using your knowledge of the criterion for uv absorption indicate which of the products (A,B,C,D) give rise to the absorption at 233 nm. If the mean molar absorptivity of the absorbing species is 2.6×10^3 m^2mol^{-1} calculate the fraction of the original linoleate which has been converted to monohydroperoxides after 90 minutes irradiation.

Products B and C are conjugated dienes whereas products A and D are unconjugated. Using the diene rules we expect B and C to have absorption at 217 nm + 15 nm = 232 nm close to the observed position. To measure the extent of conversion we note the reaction mixture has been diluted 1:100 and that the absorbance change after 90 minutes is approximately 0.5 in a 1 cm cell. This corresponds to a concentration of $0.5/2.6 \times 10^3$, or before dilution, of 100 times that. This is equivalent to the formation of about 0.002 mol dm^{-3} of conjugated species. If we assume the conjugated and unconjugated species are formed in equal amounts, then the total change is about 0.004 mol dm^{-3} product or about 10% of the original ester.

The above illustrates the way in which uv spectroscopy can be used as an aid to following a reaction both mechanistically and quantitatively.

4.6.2 Following Fast Reactions by Kinetic Spectroscopy

A number of fast reaction techniques utilise changes in uv/visible absorption as the means of following the formation and/or reaction of short-lived or transient species. Some common examples are stopped-flow and pulsed irradiation techniques.

In these systems lasers are employed for very short flash irradiations and the formation and decay of various species mentioned at fixed uv wavelengths. From the decay curves obtained the kinetics of the process can be postulated.

However, the procedures and details required for these studies is outside the scope of this ACOL Unit. But by now you should be well aware that uv/visible spectrometry is a very valuable tool for the analyst in all forms of chemical and biochemical studies for both qualitative and quantitative determinations.

Summary

The wavelengths and intensity of the ultraviolet-visible radiation absorbed by a substance is related to chemical structure, and the spectra of organic compounds are particularly sensitive to the presence of unsaturation and elements with lone-pair electrons. By assessing the contributions due to various structural features and functional groups of an organic molecule it is possible to calculate (with a high degree of success) the wavelength at which many substances exhibit their peak absorption.

Objectives

You should now:

- be aware that different chemical compounds can have an apparent identical colour but different absorption spectra;
- have learnt that well-defined chemical features have a major influence on the nature of the uv/visible spectrum;
- be able to carry out simple summations to calculate the wavelength of maximum absorption for compounds;
- understand the importance of special measurements in studying chemical reactions;
- have a knowledge of the value of uv/visible spectrometry in qualitative and quantitative analysis.

Self Assessment Questions and Responses

SAQ 1.1a

Wavenumber ($\bar{\nu}$) values in cm^{-1} units are calculated by taking the reciprocal of the wavelength (λ) values, and multiplying by an appropriate factor to allow for the conversion of units.

(*i*) What is the relationship between wavelength values in nm and wavenumbers in cm^{-1} units?

(*ii*) Similarly what is the relationship between wavenumber values in cm^{-1} units and frequency values in units of hertz (or s^{-1})?

⟶

SAQ 1.1a (cont.)

Use these two relationships to calculate the wavenumber and the frequency of yellow radiation of wavelength 575 nm. Check your values against the wavenumber and frequency scales of Fig. 1.1a.

Response

(*i*) Wavenumber = 1/wavelength or $\bar{\nu} = 1/\lambda$

Allowing for units

Wavenumber (cm^{-1}) = 10^7/wavelength(nm)

(*ii*) $\bar{\nu} = 1/\lambda = \nu/c$

Wavenumber = Frequency/ Velocity (c)

Wavenumber (cm^{-1}) = Frequency(hertz) × 100/Velocity (ms^{-1}) or Frequency = Wavenumber × Velocity/100

Applying these to yellow radiation of wavelength 575 nm

$\bar{\nu} = 10^7/575 = 17391\ cm^{-1} = 17391 \times 10^2\ m^{-1}$

$\nu = 17391 \times 10^2 \times 3 \times 10^8 = 5.2 \times 10^{14}\ s^{-1}$ (hertz)

These values agree with the scales in Fig. 1.1a.

****

SAQ 1.1b

Using your knowledge of the colours of the common reagent solutions listed below, identify the solutions corresponding to the spectra A to E in Fig. 1.1d. ⟶

SAQ 1.1b (cont.)

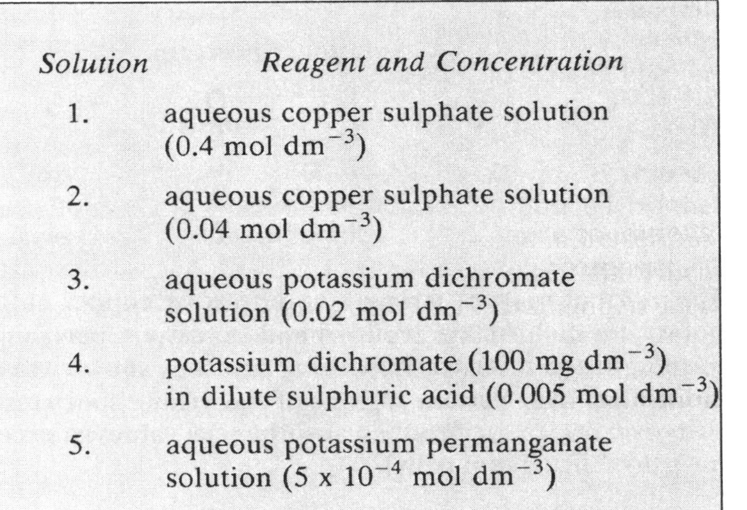

Solution	*Reagent and Concentration*
1.	aqueous copper sulphate solution (0.4 mol dm^{-3})
2.	aqueous copper sulphate solution (0.04 mol dm^{-3})
3.	aqueous potassium dichromate solution (0.02 mol dm^{-3})
4.	potassium dichromate (100 mg dm^{-3}) in dilute sulphuric acid (0.005 mol dm^{-3})
5.	aqueous potassium permanganate solution (5×10^{-4} mol dm^{-3})

Comment, if possible, on the relative intensities of absorption (relative absorptivities) of the three compounds.

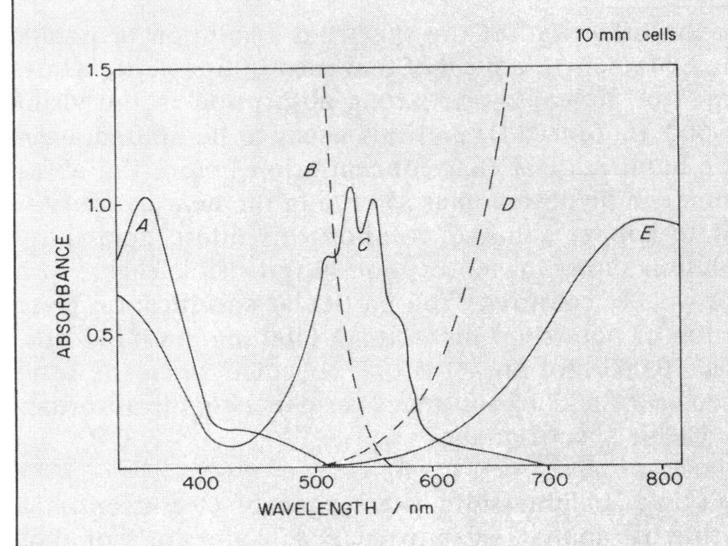

Fig. 1.1d. *uv/visible spectra of common reagents*

Response

Solution	Spectrum
1.	D
2.	E
3.	B
4.	A
5.	C

The normal reagent strength solutions of copper sulphate (blue), potassium dichromate (yellow) and potassium permanganate (deep purple) are so intensely absorbing that they show virtually complete absorption over certain regions of the visible spectrum (ie they absorb over 99.9% or produce absorbances values in excess of 3 when measured in 10 mm cells).

Thus 0.4 mol dm^{-3} copper sulphate solution (approximately 10% w/v) absorbs strongly above 640 nm (curve D) but when diluted to one-tenth the concentration absorption maximum is seen to lie at 780 nm in the near ir! (curve E).

At the other end of the spectrum a solution of potassium dichromate of strength typical of that used in titrimetric analysis (0.02 mol dm^{-3} or above) shows strong absorption in the visible region up to 500 nm (curve B) and this needs to be diluted approximately to one-hundredth of that concentration before the absorption maximum can be observed at 350 nm in the near uv (curve A). Like the dilute copper sulphate solution this dilute potassium dichromate solution shows low absorption in the visible region and hence both are weakly coloured. The particular specification given for the solution of potassium dichromate (100 mg dm^{-3}) in dilute sulphuric acid (0.005 mol dm^{-3}) is that specified when the solution is to be used as a standard substance for checking the absorbance scale of a uv/visible spectrometer.

Solutions of potassium permanganate at concentrations used for titrimetric analysis (also typically 0.02 mol dm^{-3} or above) show virtually complete absorption across the whole visible spectrum with only narrow transmission bands at 420 nm and 760 nm (not shown) producing the deep purple colour observed. Such strong solutions also need to be diluted to one-hundredth of that concentration in order to give the measurable absorption maximum at 530 nm shown in curve C. The multiple peak appearance of curve C is a unique characteristic of the permanganate ion absorption (see Part 4 of the present Unit).

From the above figure it is possible to read off approximate values of the absorbances of the three solutions at the wavelengths of maximum absorption and to calculate the molar absorptivities, using the Beer-Lambert Law expression. You could try these calculations to see if you obtain answers close to the literature values noted below. Don't worry if you can't we will be covering this in a later section.

$CuSO_4$ 25 $dm^3\ mol^{-1}\ cm^{-1}$ at 780 nm.

$K_2Cr_2O_7$ 250 $dm^3\ mol^{-1}\ cm^{-1}$ at 351 nm.

$KMnO_4$ 2300 $dm^3\ mol^{-1}\ cm^{-1}$ at 523 nm.

****

SAQ 1.2a

Carry out the following calculations:

(*i*) Obtain a value for the absorbance of a solution which only transmits 12% of the incident light.

(*ii*) Calculate the percentage of light transmitted for a solution with an absorbance value of 0.55.

(*iii*) Determine the value for the absorbance of a solution of an organic dye (0.0007 mol dm^{-3}) in a cell with a 2 cm path length if its absorptivity is 650 $dm^3\ mol^{-1}\ cm^{-1}$.

Response

In all cases it is a matter of placing the numerical values for the appropriate terms of the equations:

$$A = \log \frac{1}{T} = \log \frac{I_o}{I} = acl$$

(*i*)

$$A = \log \frac{100}{12} = \log 8.333$$

$$= 0.92$$

(*ii*)

$$0.55 = \log \frac{100}{I}$$

$$\log 3.548 = \log \frac{100}{I}$$

$$I = \frac{100}{3.548}$$

$$I = 28\%$$

(*iii*)

$$A = 650 \times 0.00070 \times 2$$

$$A = 0.91$$

If you have followed the steps in the development of the equations you should have found this fairly straightforward at this stage.

****

SAQ 1.2b

Calculate the concentration, in units of mg dm^{-3}, of a solution of each of the two compounds A and B.

Compound	M_r	$\varepsilon/dm^3\ mol^{-1}\ cm^{-1}$	Absorbance
A	250	1000	0.10
B	250	100 000	0.10

What is the significance of the molar absorptivity in analysis?

Response

Rearrange Eq 1.7 $\varepsilon = A/cl$

$$c = A/\varepsilon l$$

Compound A:

$$c = 0.10/1000 \times 1$$
$$= 1 \times 10^{-4}\ \text{mol dm}^{-3}$$
$$= 250 \times 10^{-4}\ \text{g dm}^{-3}$$
$$c = 25.0\ \text{mg dm}^{-3}$$

Compound B:

$$c = 0.10/100\,000 \times 1$$
$$= 1 \times 10^{-6}\ \text{mol dm}^{-3}$$
$$= 250 \times 10^{-6}\ \text{g dm}^{-3}$$
$$c = 25 \times 10^{-2}\ \text{mg dm}^{-3}$$

Clearly the greater the absorptivity the lower the concentration of analyte needed to give a measureable signal. You won't be surprised to learn that β-carotene, having a molar absorptivity similar in magnitude to compound B, is highly coloured and only small concentrations are needed to give the intense yellow-orange colour of carrots.

****

SAQ 1.3a

Fig. 1.3d below shows the optical components and layout of a typical recording spectrometer designed to operate over the wavelength range 190 to 900 nm. Identify on the diagram the four principal components:

Source

Monochromator

Sampling area

Detector

When the instrument is operated over certain wavelength ranges filters are inserted into the optical path. Specify which of the filters, red or blue, is used:

(*i*) at 780 nm

(*ii*) at 390 nm

In each case briefly explain the function of the filter used. What special precautions would you adopt to ensure the optimum instrument performance when measurements are being taken at 195 nm? ⟶

SAQ 1.3a (cont.)

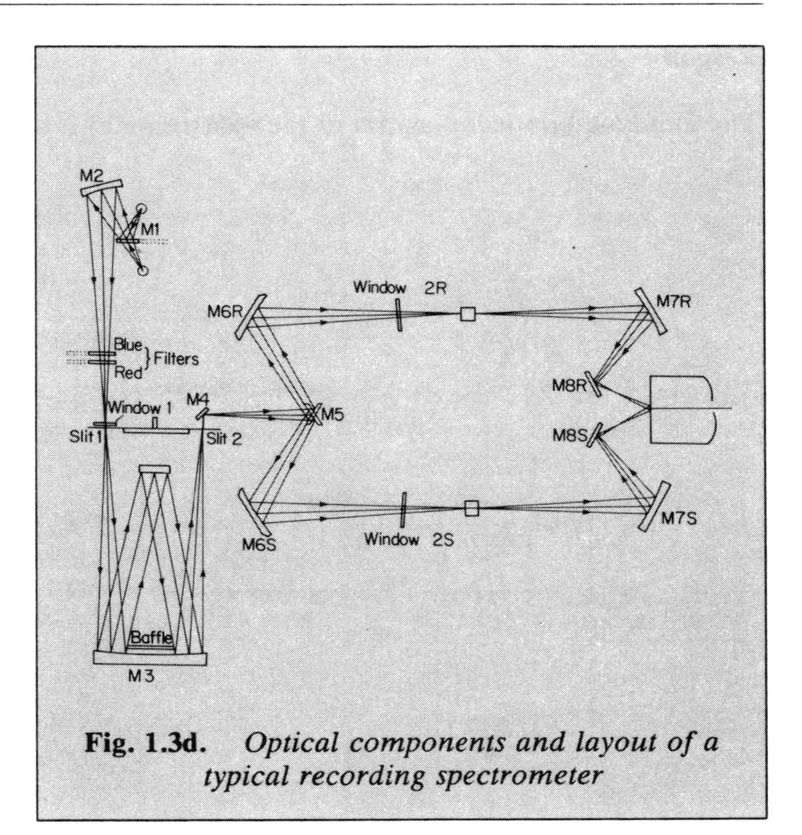

Fig. 1.3d. *Optical components and layout of a typical recording spectrometer*

Response

The annotated optical diagram of the spectrometer is shown below.

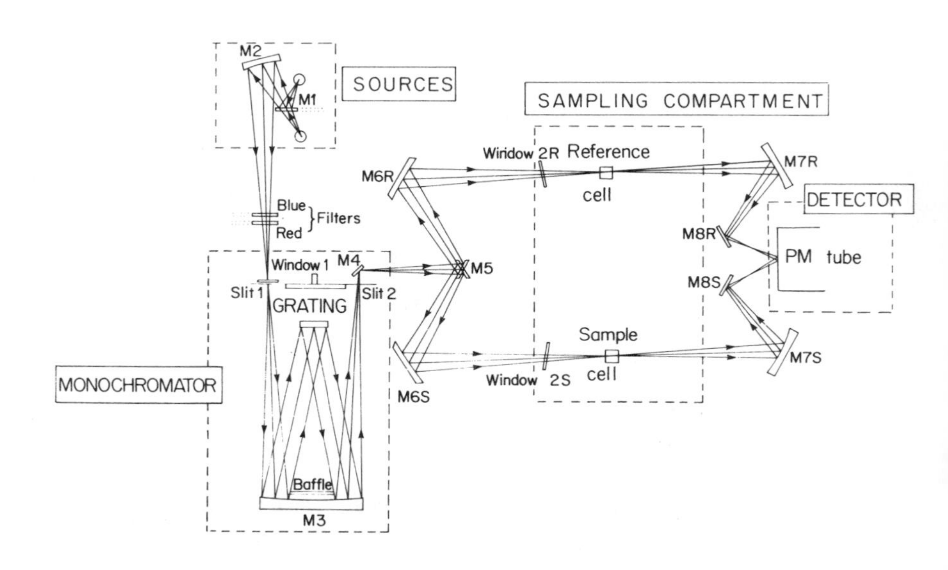

Fig. 1.3d. *Optical components and layout of a typical recording spectrometer*

You should have had little difficulty in indicating the four main components. Note that with a uv/visible spectrometer the use of two sources to cover the range is normal. A tungsten or tungsten-halogen source is usual for the range 330 to 900 nm (and beyond), although the precise change-over point in the ultraviolet can often be varied over the range 325 to 385 nm. At lower wavelengths a deuterium lamp is normally used.

At 780 nm the red filter would be used to eliminate second order wavelengths (eg 390 nm) from passing through the grating monochromator.

At 390 nm the blue/violet filter would be used to improve the stray light performance by filtering out most of the visible range from the tungsten source.

When operating at 195 nm it is necessary to purge the instrument with nitrogen as the oxygen in the atmosphere absorbs at this wavelength. In addition, cells and solvents have to be carefully chosen to ensure adequate transmittance at this low wavelength. In the absence of a double monochromator, stray radiation effects will be significant at this wavelength. (see Part 2 of this Unit).

****

SAQ 1.4a

Indicate which of the following statements are *true* and which are *false.*

(*i*) Quantitative colorimetric analysis requires the use of a light measuring instrument.

(*ii*) The eye is as good as an instrument for detecting colour changes.

(*iii*) All colorimetric methods of analysis have been developed for trace levels. ⟶

SAQ 1.4a (cont.)

(*iv*) Certain biochemical methods of analyses can be designed to be highly selective, even though they all involve measurement of change in the uv spectrum of NAD (nicotinamide-adenine dinucleotide).

(*v*) The use of uv/visible spectrometry in the monitoring of the separation of mixtures by high performance liquid chromatography (hplc) is limited to those components which show strong absorption above 220 nm.

Response

(*i*) – FALSE

(*ii*) – TRUE

(*iii*) – FALSE

(*iv*) – TRUE

(*v*) – FALSE

(*i*) The visual comparator methods described in Section 1.2 use the eye as the device for comparing colour intensities, and hence performing quantitative colorimetric analyses. However, such a colour comparison can also be made with the simplest of equipment – a pair of test tubes to hold the coloured solutions.

(*ii*) This is true since the eye is particularly good at detecting small colour differences. Instruments capable of competing with the eye in this task are modern sophisticated recording spectrometers of high precision.

(*iii*) Although many colorimetric methods have been designed for the analysis of absorbing species at low levels of concentration (typically 10^{-2} to 10^{-5} mol dm^{-3}), the higher levels of this range could not be classified as trace levels. In addition many spectrometric methods of analysis involve significant dilutions for systems containing the components to be analysed at up to molar levels of concentration.

(*iv*) Biochemical analyses utilising NAD or its phosphate derivative NADP normally involve an enzymatic species which is specific for the component being analysed. The change in the uv spectrum occurs typically as a secondary but quantitative reduction/oxidation reaction of the NAD or NADP with the products of the enzyme conversion (an example is given in Part 3 of the present Unit).

(*v*) This is false since it is now common practice to label chemically, with uv absorbing or fluorescing groups, the components to be separated. In addition the performance of uv monitors has been improved in the low uv range to enhance the range of applications.

****

SAQ 2.1a

A biochemical enzymatic analysis is being carried out at 340 nm by spectrometric measurements. Indicate which of the following would result in a large (L) and which would result in a small (S) effect on the measured absorbance.

(*i*) The sample becomes cloudy due to poor solubility. L/S

⟶

SAQ 2.1a (cont.)

(*ii*) The sample is accidently placed in a glass cell instead of a silica cell. L/S

(*iii*) The sample cell is accidently contaminated with propanone. L/S

(*iv*) The tungsten source is used instead of the deuterium source. L/S

(*v*) The pH of the reaction system is not adjusted to the optimum value. L/S

Response

(*i*) L. Cloudiness results in significant scatter of uv radiation, and hence a significant increase in apparent absorbance is expected.

(*ii*) S. This would probably have a small effect because 340 nm is significantly above the glass cut-off which occurs at about 310 nm. However, the effect with plastic cells would be much greater due to their higher cut-off wavelength.

(*iii*) L. Propanone would have a large effect since 340 nm is near the cut-off wavelength of 331 nm quoted in Fig. 2.1a. Also an organic solvent like propanone may influence an enzyme reaction.

(*iv*) S. 340 nm is within the acceptable changeover range between the sources, although stray radiation effects may be more significant with the tungsten source (see Section 2.4).

(*v*) L. Although this topic has not been specifically mentioned your general knowledge of enzyme behaviour should be sufficient for you to appreciate the importance of pH control in this type of analysis (see Section 3.1), and optimum conditions are essential for correct analyses.

****

SAQ 2.2a

Potassium thiocyanate and 1,10-phenanthroline have both been used as reagents for the determination of low concentrations of iron. Both have advantages and disadvantages for this application. Assign as many as possible of the following advantages and disadvantages to the two reagents.

Advantages

(*i*) Complex formation requires only the addition of the reagent and some acid.

(*ii*) The reagent is cheap.

(*iii*) The complex is stable and relatively free from interferences.

(*iv*) The molar absorptivity is over 1000 m^2 mol^{-1}.

(*v*) It is applicable to iron in the Fe(III) state.

Disadvantages

(*vi*) The iron must be reduced to the Fe(II) state.

(*vii*) The complex is non-stoichiometric. ⟶

SAQ 2.2a (cont.)

(*viii*) The molar absorptivity is too low to be chosen for water analysis (in the UK).

(*ix*) pH control is important.

(*x*) The complex is sensitive to light, and is relatively unstable.

Response

Potassium thiocyanate *i,ii,v,vii,viii,x*

1,10-phenanthroline *iii,iv,vi,viii,ix*

(see Section 2.2 for the details).

Of these you may have been unable to assign (*i*) because you did not have full data on the analytical procedure, and (*iii*) as no indication of possible interferences for 1,10-phenantroline has been given.

****

SAQ 2.3a

Calculate the absorptivities of $KMnO_4$ using the following data:

A $KMnO_4$ solution at λ_{max} = 522 nm gave an absorbance = 1.236 in a 10 mm cell.

The Mn concentration is 30 mg dm^{-3} [A_r(Mn) = 54.938]

(*i*) Molar Absorptivity, ε_{max}

(*ii*) Absorptivity, $E_{1\%}^{1cm}$,

Response

(*i*) The molar concentration of $KMnO_4$ = $30 \times 10^{-3}/54.938$ mol dm^{-3} [1 mole $KMnO_4$ contains 1 mole Mn]

$$= 546 \times 10^{-6} \text{ mol dm}^{-3}$$

$$\varepsilon_{max} = 1.236/(546 \times 10^{-6} \times 1)$$

$$= 2264 \text{ dm}^3 \text{ mol}^{-1} \text{ cm}^{-1}$$

$$= 226 \text{ m}^2 \text{ mol}^{-1}$$

You will recall a previous calculation with slightly different data gave 230 m^2 mol^{-1}.

(*ii*) The % concentration is calculated as follows.

Mn concentration is 30×10^{-3} g dm^{-3}

$$= 3 \times 10^{-3} \text{ g per } 100 \text{ cm}^3 \text{ (\% w/v)}$$

$$KMnO_4 \text{ concentration} = 3 \times 10^{-3} \times 158.032/54.938$$

$$= 8.63 \times 10^{-3} \text{ \% w/v}$$

$$E_{1\%}^{1cm,} = 1.236/(8.63 \times 10^{-3} \times 1)$$

$$= 143 \text{ cm}^2 \text{ g}^{-1}$$

****

SAQ 2.3b

> A solution of potassium permanganate at a manganese concentration of 3.4 mg dm^{-3} Mn transmits 23% at 522 nm and 57.5% at 480 nm. Calculate the effect of a +1% transmittance error on the absorbance at these two wavelengths. Indicate, with reasons, which wavelength is best used for the analysis of permanganate solutions. (522 nm is the λ_{max} for $KMnO_4$, 480 nm is on the side of the absorption band).

Response

At λ = 522 nm

Absorbance values:

$$A = \log 100/23$$

$$= 0.638$$

$$A = \log 100/24$$

$$= 0.620$$

$$\Delta A = 0.018$$

At λ = 480 nm

$$A = \log 100/57.5$$

$$= 0.240$$

$$A = \log 100/58.5$$

$$= 0.233$$

$$\Delta A = 0.007$$

%Absorbance error respectively;

$$0.018/0.638 = 2.8\%.$$

$$0.007/0.240 = 2.9\%$$

Thus percentage errors are comparable, under the above conditions. Measurements are best taken at 522 nm for following reasons;

1. Errors due to wavelength shifts are minimised.

2. Calibration graphs have higher slope, hence the concentration readout errors are minimised.

3. Calibration graphs show less scatter at λ_{max} values.

****

SAQ 2.4a

> List two of the problems that occur when calibration data do not obey the Beer-Lambert Law over the concentration range of interest.

Response

— Drawing an accurate curve.

— Interpolating accurately in the non-linear regions.

— Uncertainty about the behaviour of sample solutions whose composition may be quite different from those of the standard solutions. It is possible, of course, for standard solutions to give a linear Beer-Lambert relationships and for the sample solutions a non-linear relationship due to the presence of non-analyte components arising from the sample matrix itself.

****

SAQ 2.4b

Absorbance measurements are to be made on a series of aqueous solutions of copper sulphate (0 to 0.1 mol dm^{-3}, λ_{max} = 800 nm, ε_{max} = 20 m^2 mol^{-1}). The instrument being used covers the region 350 to 900 nm and has a stray light specification of 1%.

Unfortunately the detector has low sensitivity at wavelengths above 700 nm. Indicate what each of the following options has to offer in our search for high precision and simplicity of analysis.

(*i*) Increase instrument bandpass from 2 to 8 nm to improve signal strength,

(*ii*) Decrease the sample path length from 10 to 5 mm in order to diminish the effects of stray radiation,

(*iii*) Modify the analysis procedure and form the copper-ammonia complex with λ_{max} = 600 nm and ε_{max} = 80 m^2 mol^{-1},

(*iv*) Modify the analysis procedure and form the Cu-DEDC complex with λ_{max} = 430 nm and $\varepsilon_{max} \approx 10^4$ m^2 mol^{-1}.

Response

This is quite a demanding question, but you should have been able to make a lot of useful comments.

(*i*) Since copper sulphate solution has a broad absorption band increasing the bandwidth would not give rise to problems of non-linear calibration. The signal strength would increase and improve precision but it is unlikely that it would improve substantially. The only way to find out is to try it, and this is very easy to do.

(*ii*) A calculation based on a 0.1 mol dm^{-3} solution in a 10 mm cell shows that an absorbance of about 2 is to be expected. If you can recall the effects of stray radiation (see Fig. 2.4b and 2.4c) you will realise that we will be making measurements in a region of severe deviations from the Beer-Lambert Law. Reducing the path length to 5 mm will bring the absorbance down to about 1 which is satisfactory as far as the non-linearity caused by stray light is concerned. It does not however, help to improve the low precision due to the lack of detector sensitivity at 800 nm.

(*iii*) the copper-ammonia complex is readily formed, and sample dilution to bring absorbance values below 1 is straightforward. The detector is sensitive at λ_{max} = 600 nm. This looks to be a satisfactory option.

(*iv*) The sensitivity problem is well and truly overcome, but you will recall from Part 1 that the use of DEDC requires solvent extraction. The simplicity of option (*iii*) may be preferred.

****

SAQ 3.1a

In many colorimetric analyses it is sufficient to correct the sample absorbance (S) for the blank reading (B), and to read off the component concentration from the Beer's Law calibration graph. This is implied in the above procedure for iron in water when the apparent absorbance is given by:

$$R = S - B$$

Indicate whether using the above simple procedure the factors below would result in a high (H), low (L) or correct (C) value for the iron content of the water being analysed. If you have insufficient information to make a judgement use the code (I).

(*i*) The temperature of the sample dropped to 15 °C before measurement.

(*ii*) The instrument was found to have a 1 nm calibration error.

(*iii*) The sulphuric acid reagent was found to contain 1 mg dm^{-3} iron.

(*iv*) The deionised water used for the blank was found to contain 0.1 mg dm^{-3} iron.

(*v*) Only 1/10th of the quantity of hydroxylammonium chloride reagent was added (0.2 cm^3 instead of 2.0 cm^3).

(*vi*) The final solution for measurement looked slightly cloudy.

Response

(*i*) C. The TPTZ method for iron appears to be relatively temperature insensitive since the procedure merely requires the temperature to be adjusted to somewhere between 15 and 30 ^{o}C. For certain procedures a low temperature may reduce the rate of achieving a stable reading.

(*ii*) I(C). Coloured complexes usually have fairly broad absorption maxima and therefore an error of 1 nm is unlikely to affect the measured absorbance. However, to be sure we would need to have information about the shape of the absorption curve in the vicinity of λ_{max}.

(*iii*) C. A small amount of iron in the reagents would be compensated for by the blank reading.

(*iv*) L. However, iron in the water used in the blank would result in a low value of the iron and this is allowed for in Section 9.7 and 9.8 of the procedure and by the quantity C_w in the calculation of the final result in 9.11.

(*v*) L(I). Small variations in the amounts of the complexing reagents are unlikely to affect the results as normally excess reagent is added. However, adding only 10% of the required amount is likely to be insufficient to complex all the iron in the water and this is confirmed by consultation of the original report on the development of the analytical method in *The Determination of Iron in Water* by W.K. Dougan and A.L. Wilson, The Water Research Association, Marlow, England, June 1972.

(*vi*) H. Cloudiness in the final solution probably indicates turbidity and hence light scattering leading to a higher than expected absorbance of the analytical light beam. The procedure 9.6 allows for the correction S_1 for scattering and

$$R = S - B - S_1$$

If the scattering is due to suspended organic matter this should be eliminated by a wet-oxidation pretreatment which is detailed in Section 8 of the method.

****

SAQ 3.1b

The success of the enzymatic determination of glucose by the method described in Section 3.1.2 is implicit in the statement that 'NADPH is stoichiometric with the amount of glucose and is determined by means of its absorption at 334, 340 or 365 nm'. Why do you think three wavelengths are specified and what are the implications in terms of the instrumental precision achievable?

Response

Three wavelengths are specified to allow the use of instruments which employ filtered mercury light sources as well as those which employ the normal tungsten light source with a variable wavelength monochromater. With the latter the wavelength would be set for the λ_{max} of NADPH ie 340 nm. With such instruments we normally avoid using the sloping side of an absorption band which is subject to wavelength setting and bandwidth errors (Section 2.4). A mercury line source does not suffer from wavelength errors (only focussing errors) but can have some other drawbacks in terms of adaptability to other types of analysis.

The method of calculation gives values of ε appropriate to the three wavelengths, but these ε values may not be fully reproducible on different uv instruments!

****

SAQ 3.1c

The molar absorptivity (ε_{max}) for the ternary complex of iron(II), 4-chloro-2-nitrosophenol and Rhodamine B is quoted as 9.0×10^3 m^2mol^{-1} at 558 nm. This is only a factor of four (4×) times greater than that obtained with the iron(II), tripyridyltriazine (TPTZ) complex, yet the typical iron concentration levels measured are at least ten times lower (50 $\mu g\ dm^{-3}$ against 500 $\mu g\ dm^{-3}$). How is this achieved?

Response

The principal reason is that the complex is extracted from 20 cm^3 aqueous solution into 5 cm^3 toluene prior to measurement. However, this further four-fold improvement is eliminated by the choice of a 10 mm path length for the ternary complex in the toluene as against a 40 mm path length for the TPTZ method.

The minimum detectable levels quoted for the TPTZ method is 3 to 15 $\mu g\ dm^{-3}$ Fe(II) corresponding to an absorbance of only 0.004 to 0.02. The lowest absorbance quoted in the Analyst report (not given here) is 0.042 ± 0.001 which corresponds to 0.13 ± 0.003 $\mu g\ dm^{-3}$. Such precision in absorbance measurements suggests that a modern instrument of high precision was used in the latter investigation.

****

SAQ 3.3a

A certain dye A has the same absorption spectrum in aqueous solution containing either 10% pyridine or 10% ethanol. Dye B behaves in the same way. A mixture of the two dyes, however, does not give the same absorption spectrum in the two solvents. Determine from the data below the composition of the mixture, showing how the more satisfactory data are selected.

	450 nm	550 nm	650 nm
Dye A (1 g dm^{-3})	1.75	0.68	0.21
Dye B (1 g dm^{-3})	0.11	0.23	0.33
Mixture in ethanol-water	2.07	1.75	1.30
Mixture in pyridine-water	2.08	1.37	1.20

(Figures given in the tables are absorbances in a 10 mm cell at the wavelengths indicated)

To help you start, treat the problem as one involving a binary mixture, choose the two most appropriate wavelengths for the analysis, and then check that the result fits at the third wavelength. You will have to decide which solvent system to choose for the calculation - only one solvent gives satisfactory results.

Response

Treat the problem as a binary analysis at 450 nm and 650 nm and check that the result fits at 550 nm.

As it is the less polar solvent, pyridine is likely to cause less interaction than would ethanol, therefore we set up simultaneous equations as follows:

$$\text{at 450 nm } 2.08 = 1.75\ C_A + 0.11\ C_B$$

$$\text{at 650 nm } 1.20 = 0.21\ C_B + 0.33\ C_B$$

$$\text{Solving gives } C_A = 1 \text{ g dm}^{-3} \text{ and } C_B = 3\text{g dm}^{-3}$$

Check at 550 nm $0.68 + 3 \times 0.23 = 1.37$ as observed.

As the observed and calculated values agree for 550 nm additive behaviour clearly applies in aqueous pyridine.

For ethanol/water mixtures the same procedure gives

$$2.07 = 1.75\ C_A + 0.11\ C_B$$

$$1.30 = 0.21\ C_A + 0.33\ C_B$$

Solving this time gives

$$C_A = 0.98 \text{ g dm}^{-3} \text{ and } C_B = 3.28 \text{ g dm}^{-3}$$

Check at 550 nm gives

$$0.68 \times 0.98 + 3.28 \times 0.23 = 1.42 \neq 1.75$$

So the observed and calculated values do not agree and the two dyes show non-additive behaviour in aqueous ethanol.

****

SAQ 4.2a

Fig. 4.2b is an absorption spectrum of potassium permanganate in which the four major peaks have been numbered. Values for λ_{max} and ε_{max} for the strongest peak are given. Measurements were made using a 1 cm cell.

Complete the table below to show the λ_{max} and ε_{max} values of the three major bands

	λ_{max}/nm	ε_{max}/m^{2}mol^{-1}
Band 1	533	1.01×10^3
Band 2		
Band 3		
Band 4		

→

SAQ 4.2a (cont.)

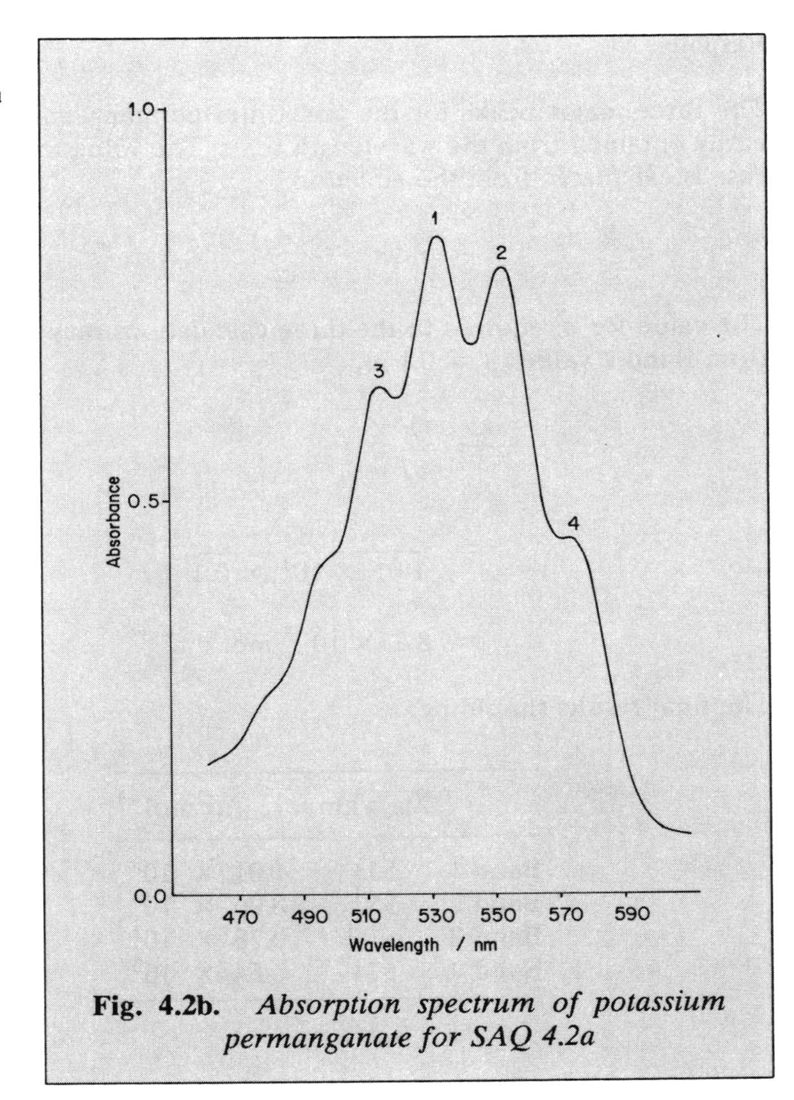

Fig. 4.2b. *Absorption spectrum of potassium permanganate for SAQ 4.2a*

Response

The three major peaks for the potassium permanganate curve are easily obtained from the wavelength scale. The value for ε_λ in each case is calculated from the equation:

$$\varepsilon = \frac{A}{cl}$$

The value for c required in the three calculations may be obtained from Band 1 values, $l = 0.1$ m.

$$c = \frac{A}{\varepsilon l}$$

$$= \frac{0.83}{1.01 \times 10^3 \times 0.1}$$

$$= 8.2 \times 10^{-3} \text{ mol dm}^{-3}$$

The final results should be:

	λ_{max}/nm	ε_{max}/m^{2}mol^{-1}
Band 1	533	1.01×10^3
Band 2	552	0.96×10^3
Band 3	514	0.78×10^3
Band 4	574	0.54×10^3

****

SAQ 4.4a

Assign from list 2 below, the appropriate values of λ_{max} and ε_{max} for each system given in list 1. Comment on the shifts observed relative to benzene λ_{max} = 255 nm; ε_{max} = 23 m^2mol^{-1} in terms of the electronic properties of the auxochromes.

List 1 Absorbing systems

A aqueous solution of aniline

B solution A with added 0.01 M HCl

C aqueous solution of phenol

D solution C with added 0.01 M NaOH

List 2 Absorption Characteristics

	λ_{max}/nm	ε_{max}/m^2mol^{-1}
E	255	20
F	270	140
G	280	130
H	290	230

Response

List 1 – List 2 Combinations

A – F(G)
B – E   see below for comments on
C – G(F)   alternatives listed
D – H

Chromophore is the benzene ring in each case

Auxochromes are NH_2(A), NH_3^+ (B), OH(C), O^-(D)

The shifts relative to benzene(λ_{max} 255 nm) depend on the electron-donating property of the auxochrome, which are in the order

$$NH_3^+ << OH < NH_2 << O^-$$

This order is essentially the order which can be deduced from a knowledge of the basic chemistry of these groups in aromatic substitution. You might not have appreciated that the lone pair of electrons in the NH_2 produces a greater red (bathochromic) shift than that on the OH group and hence the alternatives given above. You should have been aware that the addition of acid to form NH_3^+ effectively removes the nitrogen lone pair, whilst the formulation of RO^- is generally associated with significant bathochromic shifts.

****

SAQ 4.5a

A compound with the following structure

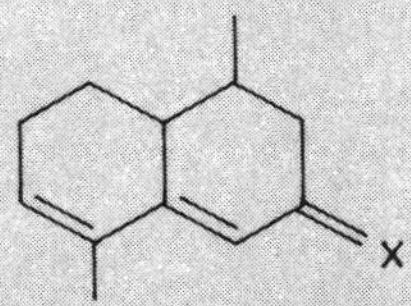

is classed as (*i*) an alkene if X is CH_2,

and (*ii*) an unsaturated ketone if X is O.

For the purposes of applying the Woodward rules to predict the λ_{max} of the uv spectrum in alcohol solution we need to be able to characterise aspects of the structural contributions.

(A) In this context indicate which of the following statements are *true* and which are *false*.

(*i*) When X is CH_2 we have a triene and the diene rules do not apply.

(*ii*) The compound has only *two* exocyclic double bonds.

(*iii*) The compound has two methyl groups both of which contribute +5 nm to the absorption.

(*iv*) When X is an oxygen atom (O) the double bonds of the unsaturation are in the α, β and γ, δ positions.

(B) Using the appropriate rules, show that the predicted λ_{max} (EtOH):

(*i*) for the alkene structure is 269 nm,

(*ii*) for the ketone structure is 277 nm.

Response

(A)

(*i*) FALSE, although this *is* a triene the diene rules *do* apply.

(*ii*) FALSE, there is only *one* exocyclic double bond, the centre one of the triene.

(*iii*) FALSE, only *one* methyl group contributes to the system, although there are also *three* ring residues to take into consideration.

(*iv*) TRUE, the basic system is:

$$\diagdown \overset{\delta}{CH} = \overset{\gamma}{C} - \underset{\beta}{C} = \underset{\alpha}{CH} - C = O$$

(B) The wavelengths are obtained as follows

	(*i*)	(*ii*)
Acyclic diene	214	
α, β unsaturated carbonyl		222
Extended conjugation	30	30
Exocyclic Double Bond	5	5
Methyl Group	5	5
Ring Residues (3)	15	15
Total	269 nm	277 nm

****

Units of Measurement

For historic reasons a number of different units of measurement have evolved to express quantity of the same thing. In the 1960s, many international scientific bodies recommended the standardisation of names and symbols and the adoption universally of a coherent set of units—the SI units (Système Internationale d'Unités)—based on the definition of five basic units: metre (m); kilogram (kg); second (s); ampere (A); mole (mol); and candela (cd).

The earlier literature references and some of the older text books, naturally use the older units. Even now many practicing scientists have not adopted the SI unit as their working unit. It is therefore necessary to know of the older units and be able to interconvert with SI units.

In this series of texts SI units are used as standard practice. However in areas of activity where their use has not become general practice, eg biologically based laboratories, the earlier defined units are used. This is explained in the study guide to each unit.

Table 1 shows some symbols and abbreviations commonly used in analytical chemistry. Table 2 shows some of the alternative methods for expressing the values of physical quantities and the relationship to the value in SI units.

More details and definition of other units may be found in the *Manual of Symbols and Terminology for Physicochemical Quantities and Units*, Whiffen, 1979, Pergamon Press.

Table 1 *Symbols and Abbreviations Commonly used in Analytical Chemistry*

Å	Angstrom
A_r(X)	relative atomic mass of X
A	ampere
E or U	energy
G	Gibbs free energy (function)
H	enthalpy
J	joule
K	kelvin (273.15 + t °C)
K	equilibrium constant (with subscripts p, c, therm etc.)
K_a,K_b	acid and base ionisation constants
M_r(X)	relative molecular mass of X
N	newton (SI unit of force)
P	total pressure
s	standard deviation
T	temperature/K
V	volume
V	volt (J A^{-1} s^{-1})
a, a(A)	activity, activity of A
c	concentration/ mol dm^{-3}
e	electron
g	gramme
i	current
s	second
t	temperature / °C
bp	boiling point
fp	freezing point
mp	melting point
$\approx$	approximately equal to
$<$	less than
$>$	greater than
e, exp(x)	exponential of x
ln x	natural logarithm of x; ln x = 2.303 log x
log x	common logarithm of x to base 10

Table 2 *Alternative Methods of Expressing Various Physical Quantities*

1. **Mass (SI unit : kg)**

$$g = 10^{-3}\ kg$$
$$mg = 10^{-3}\ g = 10^{-6}\ kg$$
$$\mu g = 10^{-6}\ g = 10^{-9}\ kg$$

2. **Length (SI unit : m)**

$$cm = 10^{-2}\ m$$
$$Å = 10^{-10}\ m$$
$$nm = 10^{-9}\ m = 10Å$$
$$pm = 10^{-12}\ m = 10^{-2}\ Å$$

3. **Volume (SI unit : m^3)**

$$l = dm^3 = 10^{-3}\ m^3$$
$$ml = cm^3 = 10^{-6}\ m^3$$
$$\mu l = 10^{-3}\ cm^3$$

4. **Concentration (SI units : mol m^{-3})**

$$M = mol\ l^{-1} = mol\ dm^{-3} = 10^3\ mol\ m^{-3}$$
$$mg\ l^{-1} = \mu g\ cm^{-3} = ppm = 10^{-3}\ g\ dm^{-3}$$
$$\mu g\ g^{-1} = ppm = 10^{-6}\ g\ g^{-1}$$
$$ng\ cm^{-3} = 10^{-6}\ g\ dm^{-3}$$
$$ng\ dm^{-3} = pg\ cm^{-3}$$
$$pg\ g^{-1} = ppb = 10^{-12}\ g\ g^{-1}$$
$$mg\% = 10^{-2}\ g\ dm^{-3}$$
$$\mu g\% = 10^{-5}\ g\ dm^{-3}$$

5. **Pressure (SI unit : N m^{-2} = kg m^{-1} s^{-2})**

$$Pa = Nm^{-2}$$
$$atmos = 101\ 325\ N\ m^{-2}$$
$$bar = 10^5\ N\ m^{-2}$$
$$torr = mmHg = 133.322\ N\ m^{-2}$$

6. **Energy (SI unit : J = kg m^2 s^{-2})**

$$cal = 4.184\ J$$
$$erg = 10^{-7}\ J$$
$$eV = 1.602 \times 10^{-19}\ J$$

Table 3 *Prefixes for SI Units*

Fraction	Prefix	Symbol
10^{-1}	deci	d
10^{-2}	centi	c
10^{-3}	milli	m
10^{-6}	micro	μ
10^{-9}	nano	n
10^{-12}	pico	p
10^{-15}	femto	f
10^{-18}	atto	a

Multiple	Prefix	Symbol
10	deka	da
10^{2}	hecto	h
10^{3}	kilo	k
10^{6}	mega	M
10^{9}	giga	G
10^{12}	tera	T
10^{15}	peta	P
10^{18}	exa	E

Table 4 *Recommended Values of Physical Constants*

Physical constant	Symbol	Value
acceleration due to gravity	g	$9.81\ m\ s^{-2}$
Avogadro constant	N_A	$6.022\ 05 \times 10^{23}\ mol^{-1}$
Boltzmann constant	k	$1.380\ 66 \times 10^{-23}\ J\ K^{-1}$
charge to mass ratio	e/m	$1.758\ 796 \times 10^{11}\ C\ kg^{-1}$
electronic charge	e	$1.602\ 19 \times 10^{-19}\ C$
Faraday constant	F	$9.648\ 46 \times 10^{4}\ C\ mol^{-1}$
gas constant	R	$8.314\ J\ K^{-1}\ mol^{-1}$
'ice-point' temperature	T_{ice}	273.150 K exactly
molar volume of ideal gas (stp)	V_m	$2.241\ 38 \times 10^{-2}\ m^3\ mol^{-1}$
permittivity of a vacuum	ϵ_o	$8.854\ 188 \times 10^{-12}\ kg^{-1}\ m^{-3}\ s^4\ A^2\ (F\ m^{-1})$
Planck constant	h	$6.626\ 2 \times 10^{-34}\ J\ s$
standard atmosphere pressure	p	$101\ 325\ N\ m^{-2}$ exactly
atomic mass unit	m_u	$1.660\ 566 \times 10^{-27}\ kg$
speed of light in a vacuum	c	$2.997\ 925 \times 10^{8}\ m\ s^{-1}$